Praise for SWEETNESS & LIGHT

"An evocative work that feels, smells, and tastes like everything to do with bees. Readers will be absorbed into the bee yard with her as she explores images and sensations of beekeeping with all senses alert." —Mark Winston, *New York Sun*

"*Sweetness & Light* is a refreshing toast to the honeybee." —*Washington Post*

"Hattie Ellis's *Sweetness & Light* flows as reassuringly as the low buzz from a contented hive. . . . It's an intriguing journey, with fine descriptive detours." —*Saveur*

"Entrancing anecdotes, accurate details, and meticulous research add up to a sweetly satisfying read." —*Publishers Weekly,* starred review

"Those with a bent for natural history will find Ellis a class act, her style among the fanciful and insightful best. An indispensable addition to a literature already brimming with anecdote and observation." —*Kirkus Reviews*

"A marvelous combination of natural history and social science as she explores the ways of bees, honey, and humans." —*Booklist*

"Ellis is the bees' knees." —*The Guardian*

SWEETNESS

& LIGHT

The Mysterious History
of the Honeybee

HATTIE ELLIS

THREE RIVERS PRESS
NEW YORK

THREE RIVERS PRESS and the Tugboat design are registered trademarks of
Random House, Inc.

Originally published in Great Britain by Hodder & Stoughton,
London. Subsequently published in hardcover in the United States by
Harmony Books, an imprint of the Crown Publishing Group,
a division of Random House Inc., New York, in 2004.

Library of Congress Cataloging-in-Publication Data
Ellis, Hattie.
Sweetness and light : the mysterious history of the honeybee / Hattie Ellis.
1. Honeybee.   2. Bee culture.   I. Title.
QL568.A6E68   2004
595.79'9—dc22   004004116

ISBN-13: 978-1-4000-5406-0
ISBN-10: 1-4000-5406-0

Printed in the United States of America

Design by Lynne Amft

10 9 8 7 6 5 4 3 2 1

First U.S. Paperback Edition

For Roger and Margaret Ellis

"We have chosen to fill our hives with honey and wax; thus furnishing mankind with the two noblest of things, which are sweetness and light."

—JONATHAN SWIFT (1667–1745)

CONTENTS

ACKNOWLEDGMENTS

Many people shared their thoughts and stories about bees and honey with me. It would be impossible to mention them all personally—nearly everyone I spoke to on the subject had something close to their heart to say—but such experience and inspiration were regular spoonfuls of honey to my mind and meant a great deal to me. Thank you. I'd like to thank Roger Ellis, Emily Faccini, Gordon Smith, Frances Price, and Gail Vines, who generously took the time to read through drafts, and Willie Robson, of Chain Bridge Honey Farm in Northumberland (www.chainbridgehoney.co.uk), who both read through the chapters and made me want to write about the honey bee in the first place. The International Bee Research Association's specialist library in Cardiff was a great help and a magnificent resource (www.ibra.org.uk). Finally, thanks to my agent, Georginal Capel, and my exceptional editors, Richard Atkinson in the U.K., and Kim Kanner Meisner in the U.S.

HEATHER HONEY

Willie Robson drives up to his beehives on the heather moor at Hangwell Law in the north of England. As he climbs, a road movie plays on the windshield of his truck, of speeding cars and tarmac, of garages and caravan parks, giving way to a landscape of small, upland farms with scrubby slopes, populated by sheep. Curving horizons reach to the edges of the sky. Countryside is mostly air; and people, now largely urban and suburban, often idealize this sort of landscape as beautiful in its emptiness. To those who live here, and who can remember the to-and-fro of a rural workforce, this silence can feel more like an eerie absence. The land is grand here, with its heathered horizons and open distances, yet it can also lack the intimacy of use.

Beekeepers have brought their hives onto these northern moors for at least a millennium, and some still do. July brings the bonny bell heather and its rounded, ruby-purple flowers, and then the finer, paler, more common ling heather appears, lasting until the end of August. Ling heather honey, with its unique gellike texture and a room-filling fragrance, is one of the most prized in the world. In the pot, it glows fox red, often beaded with little silver bubbles.

Willie turns the truck off the road and drives along a half-track toward a belt of pines. In front of the trees is a broken line of fifty homemade hives. Painted in quiet blues, browns, and grays, they sit like orderly toadstools, squat and odd shaped, with a square roof on top of each. For two months, the bees have been coming and going between the hives and the heather moor, collecting nectar and pollen from the flowers, to feed each colony of up to fifty thousand bees. The bottom box contains the brood comb, where the insects are born and raised. The upper boxes, or "supers," hold the excess honey, saved up as a hefty gold reserve for the winter. The beekeeper is about to raid the bank.

The truck stops. Four hundred feet up, 6 miles or so from the sea, the land here feels close to the sky. On the horizon, the heather moor fades into the far blue of hills. There is a smell of pines and a hum of bees, quiet, for now. Jumping down from the truck cab, Willie takes off his hairy Borders tweed cap and zips up in his bee suit. Beekeepers say, casually, that stings are just an occupational hazard, and they get rid of them with a practiced cross between a scratch and a flick. But they all have tales, ranging from mad bravura to comic-book chases to sly observations of others' misfortune. Honeybees left alone do not sting; stinging might harm the intruder, but it also kills the bee. The twin barbed shafts dig into the skin, pump poison into human flesh, and then cannot withdraw. Instead, the sting rips the center from the abdomen so the insect straggles toward death, its insides torn out, pink and pulsing. But bees will die to protect the hive, just as they will fly ceaselessly to collect nectar and pollen so the hive's colony can live.

The apiarist's armor is a bee suit. Willie has a sort of khakigreen, nylon flying suit that zips across the body and then across the neck to close up the net-fronted hood. The legs are tucked into boots and the arms into gloves elasticized at the wrists. In his suit, he walks around like a spaceman. Boots and gloves restrict some movement, but he goes slo-mo for another reason. "You go with a quiet tread, or all hell breaks loose," he says. "It's a matter of weighing the situation. If trouble starts, you bail out."

After finding a piece of sacking among the bric-a-brac on the back of the truck, Willie lights the cloth with a match and puts it in a smoker formed like a pair of miniature bellows. The smoke can help lull the bees. They think there is an emergency, eat their fill of honey, as if ready for flight, and either because a full bee is a happy bee or because they are now less able to bend and sting, they are less aggressive. Willie takes the top off the first hive. Pffffff, Pffffff, Pffffff, goes the smoke. After a short pause, he heaves off the top box. Immediately, the weight reveals the exact extent of the haul. Honey is one and a half times heavier than water and a full box tells on your muscles. Beekeeping, in some aspects, is like fishing: some years you get next to nothing; others, you crop gold. This year everything worked; both skill and luck came together, and it is boom time. The bees were in the right place at the right time; the weather was good over the year. Willie and his family have kept bees here for more than fifty years, and he is now reaping the rewards of knowing his turf and keeping bees that are well adapted to their environment. This trip to Hangwell Law comes after a run of collecting a bumper harvest of heather honeycomb in ten days. It does not happen every year, or even often. But today, he gets 2,500 pounds of honey. Such is the drama of harvest.

The bees, in the meantime, go purposefully berserk. Zinging, small, aggressive atoms, gold in the late-afternoon sun, attack again and again from different angles, trying to find a way into the bee suit. Their persistence is unrelenting. Bees in the wild can burrow into the fur of an attacking bear, to sting the animal where it will hurt him hopping mad. In the same way, they seek the vulnerable chink in the beekeeper's second skin. A hole in the fingertip of a glove, a stray stitch on a seam will not go unpunished. You feel like a character transposed into a video game, surrounded by flying attackers, the bee suit slightly claustrophobic, limiting your vision but not the sounds, nor the sudden sight of bees flying onto the net visor, inches from your eyes. Willie says the bees can get to people mentally. "They get you on the shake," he says. "They undermine your confidence and go dab, dab, dab." When a bee stings, a banana-like odor spreads in the air, attracting other bees to sting the same spot, like sharks coming to blood pulsing through the water.

Today, there's no real trouble. As each hive is opened, the noise grows, but Willie keeps calm, steady in the eye of the storm. Some beekeepers lose bees by carelessly crushing them under boxes as they work under the pressure of time and the bee blitz. Willie knows that bees matter more than honey. He brushes insects off each box with gentle sweeps of fern and the triumph he feels at the haul is as much about the bees as anything else: "When the opportunity presents, they are up and at it," he says, with frank admiration, as he lifts a box as heavy as a block of gold.

On the back of the truck, towers of honeycomb boxes build up, with a great cloud of bees above, moving up and down and round and round as if in a demented dance of fury. Willie, still in his bee suit, but with the hood down now, drives off. He stops after a few hundred feet and waits for a few minutes, to let some of the bees following the truckload fly back to the hives. Then he starts the engine up, drives another few hundred feet and stops again. Such thrift is the equivalent of scraping out the cake mix from the bowl: the last drops matter.

Honey is, essentially, a wild food. Willie couldn't get honey without the bees; the bees couldn't make the honey without the sun shining and the rain falling, the right amount at the right times, to bring the flowers into bloom and full of nectar. It helped, this particular year, that the winds were the mild southwesterlies rather than the usual northwests with their edge of chill. Heather is a faddish flower that needs the right conditions to bloom bright. This year the purple spread to the horizons. Clouds of pollen puffed up at every step, whitening animals' muzzles. The nectar flowed and the bees massed on the flowers, drinking deep.

Willie can provide hives for the bees; he can place his hives near good plants; he can look after the bees to a certain extent. But you can hardly term a man a livestock farmer whose herd consists of seventy-five million flying, stinging, crawling, sucking, working, laying, feeding, fighting, instinctual, independent insects. Man makes use of bees, but only by respecting their nature.

WILLIE ROBSON'S FAMILY home and business is the Chain Bridge Honey Farm, 20 miles from the heather moor, on the banks of the River Tweed. At this point, the river acts as the watery border between Scotland and England. It runs broad here, with fat, smooth curves, and its force, sprung upstream from so much rain over wide moors and blank horizons, is gathered together, ready to pour into the North Sea at Berwick-on-Tweed 3 miles downstream.

The honey farm takes its name from the Chain Bridge over the Tweed, just a couple of minutes' stroll away. At the time the bridge was built, in 1820, it was the longest suspension bridge in the world. There is a heritage plaque on the Scottish side, with a cartoon of the engineer driving an overloaded carriage at the launch; a cheery Dickensian image reflecting that, in retrospect, the good times were starting to roll. Like many other parts of Northumberland and the Scottish Borders, over the centuries the land has been metaphorically crosshatched with scars from territorial conquests, celebrated in Borders ballads of robbers, or reivers, raiding land and livestock, attacks and counterattacks, all stoked up with fireside boasts. When the bridge was built, the fighting was finally over.

Now the borderless global culture invades every highway and backwater. On the red sandstone of the English side of the Chain Bridge someone has scratched: WEST SIDE, 2 PAC, referring to a notorious gangsta rapper and his territory in Los Angeles. The incongruous reference, in this quiet spot, is nicely ironic—humans playing at frontiers, just as they play CDs of guntalk. Now, alongside castles for the tourists, you just see the roofs of big, private houses poking out amid the tall trees of settled parkland, all risen with the stability of trade and commerce as their owners amassed fortunes from fishing, mining, trade, and agriculture.

There is genuine unease, however, about the future of the countryside. In the 1980s, the wealth and ways of heavy industry collapsed; now it is the turn of the land. Willie is one of the few people making an unsubsidized living from agriculture.

All the while, amid skirmishes and prosperity, the bees go where they have always gone: to the richest, longest, sweetest draft of nectar they can find, whether it is Scottish or English; agricultural oilseed rape or heather on the moorlands. When you walk by the river in early summer, a shy galaxy of wildflowers glimmers from the ground below: pinks, yellows, sky blues, mauves. Walkers shelter from the soft, cold rain under the occasional stretch of hawthorn hedge, and then continue, with the grass and flowers glittering below their feet. As I walked by the river, I watched the bees with a new interest, looking at them go to fill the honey pot, looking at the plants and wondering what sort of honey their nectar would produce.

The hawthorn used to be a fickle but exquisite source of honey, when the fields in this part of the world were parceled off between their prickly hedges. These high hedges, flowering in May and June, kept cattle in and gave the animals shelter, and they existed because the Irish shipped their stock over to be fattened by the Northumbrian farmers, who then sold them to market. No longer.

Grain gets more tempting subsidies than livestock. The hawthorn was ripped out in the sixties when the beasts went, and the bees found other flowers.

In early June, the land also used to spread white with native clover. This helped fix nitrogen in the soil to renew its fertility. Many rural households took advantage of the clover and kept bees: an agricultural worker could double his monthly income by selling honey. Most people kept bees, and primary schools had hives to teach children about beekeeping, just as they taught sewing and gardening—unimaginable now. The clover went as farmers turned to artificial fertilizers after the Second World War, made expensively using petrochemicals. Again, the bees went elsewhere.

Countryside follows market forces; or rather government grants. Great swathes of heather have been ripped out to grow crops less suited to the moorland. Now there are subsidies for putting the heather back again. The most prolific agricultural crop for honey in Britain is oilseed rape. Heavily subsidized by the European Union's common agricultural policy, it is used for animal feed, as a break crop from cereals, for cheap vegetable oil, and recently as a biodiesel. Quite apart from all this, it is a useful source of nectar for the bees in the early spring, although some connoisseurs dislike its now common presence in British honeys, and the way it granulates so quickly.

For a while, in these parts, the fields were dusky with a purply blue, as farmers grew borage for pharmaceutical companies. The bees produced a beautiful, clear honey from the flowers. But the crop proved fickle and was dropped. You can just glimpse faint smudges of blue along field edges like the ghosts of past plants.

Wild and garden flowers are the bees' ever-changing buffet, alongside the substantial banquets of agricultural crops. Willow herb, known as fireweed, flowers in midsummer and makes a sweet pale honey. The wild pink flowers spread like flames and its nickname came from the way it colonized burnt-out bomb sites. It also grows where woodland has been cut down. In the late summer and early autumn, another particular flower, balsam, quivers with bees, and pinks the banks of the Tweed so the peaty water beside them looks like warm chocolate. The flowers spread here from seeds washed down from the eighteenth- and nineteenth-century woolen mills upstream, where imported fleeces were washed white, in the process dispatching flowers from Spain and France.

This is just one patch of the planet. Wherever they are—jungle, Arctic tundra, northern forest—bees find nectar, even just the smallest of sips from tiny desert plants.

THE CHAIN BRIDGE HONEY FARM is very much a family business, run by Willie Robson with the help of his wife, Daphne, their son, Stephen, who is also a beekeeper, and their daughters, Heather and Frances. After four generations of building up the business, the Robsons now have fifteen hundred hives, which are moved around the highlands and lowlands of Northumberland and the Berwickshire, following the flowers. A local map the Robsons have pinned up on a wall shows their heather sites, and the place names have a terse beauty: Burn Castle, Rawburn, Scarlaw, Stobswood, Old Bewick, Chatton Sandyfords, Harehope, Hangwell Law. Over winter, the hives are placed in sheltered spots near domestic gardens; in the summer, near fields of crops or stretches of wildflowers.

Chain Bridge honey is regarded with pride and affection by the region. You see the plastic tubs of their heather honeycomb and jars of Tweedside Blend on the counter in local newsagents, as well as in smart gift shops and food specialists. In an age of globalization, it is a genuine regional product, and they sell as locally as possible—not least because it saves on the costs of delivery.

In the preparation room of the honey farm, Willie is cutting up honeycombs from the boxes brought down from the heather moor. He is wearing his tweed cap, its colors reflecting the soft blues, grays, and greens of this wet, wild landscape, and tendrils of gray hair break out from under its brim. The comb's miniature golden columns are stacked one on top of the other. The room is full of the smell of honey, hanging in the air, soft and warm, alongside the sound of sticky cuts as he divides the honeycomb with a bone-handled knife. In hours of cutting and talking, not a single stray drip escapes. When a wasp comes pesking nearby, Willie dispatches it with a swift, neat backhand of his knife. "Sampras, eat your heart out," he says, in passing. He cuts off a generous corner of comb, offers it to me, and takes a bit himself. A broad, strong man, he looks, suddenly, like a bear enjoying a sweet snack. Just as suddenly, in the midst of a gloomy diatribe about the state of farming or greedy businessmen pouring money into idiot ideas, his face will light up with a sun-shaft of humor, or he will walk around the honey farm trailing out notes from his choir's repertoire: "Oh the rhythm of life is a powerful beat. . . ."

As with most small independent businesses, home and work are indivisible. Willie thinks, lives, breathes, and builds the business constantly. The honey farm, in turn, has taken on many human qualities. Despite the fact the farm is deep in the countryside, people come and go constantly, dropping by to hear the news or to see what is happening with Willie's latest scheme, telling a story as they make a purchase or arranging for an item to be picked up or sent on. People—visitors, customers, workers—are greeted as a part of the natural hospitality of the place.

Willie reckons 80 percent of the profits go back into the local economy. He employs about ten people, some of them highly skilled individuals looking for a home after the collapse of small- and medium-scale agriculture. When I visited, there was much banging and tinkering going on in some new buildings they had put up: a new vintage transport exhibit was being created to attract visitors. Wandering around vehicle entrails and the beastlike bodies of old cars and trucks, Willie told me about a tractor recently shipped over from Australia by a beekeeper who was dying of cancer. He managed to bring his prized vehicle over personally in order to give it a home at the new museum. Chain Bridge Honey Farm draws people back, and I could imagine this man traveling halfway across the world to make a final, lasting connection with this place before he died.

Willie Robson's way of treating people fairly (both workers and customers—prices are kept reasonable) feels refreshingly ungreedy; it is almost old-fashioned in the modern world—and yet it works. The honey farm produces something good—very good—without cutting corners. Hardly anyone bothers to produce heather honeycomb these days: it is too much work. Yet there is a market for it, as the Robsons prove. Farmers, in the decades of subsidies—what Willie calls funny money—have not needed to market their produce and connect with the customer. In this time of change, they are suddenly vulnerable. Willie is in a better position. He has had to keep a careful eye on what people want, developing a range of cosmetic products, such as the heather-honey lip balm so fragrant you keep licking your soft lips and developing honey foods, such as a rich cake and a tangy grain mustard. Recently, he had been trying to persuade several farmers in the area to grow borage again. He would get good food for his bees, and they could sell the crop into the flourishing health food market; the plant makes an oil that has more gamma linolenic acid than evening primrose oil, which has taken the international health food market by storm for its reputed benefits for those with afflictions from high cholesterol to rheumatoid arthritis. The subsidy route is a dead end, he believes: "If you pay people for doing nothing, it will ultimately destroy them."

The skills of beekeeping itself originate from Willie's father.

Selby Robson lectured on beekeeping in the days when it was seen as an important part of the rural economy. He would tour around the north, listening as much as he talked, picking up the knowledge of other beekeepers and passing it on. Now that farmers are being urged to diversify, they lack the likes of Selby Robson, and others who have honed such skills for themselves from time-tested experience, among their ranks.

For Willie's part, he says that, increasingly, his knowledge comes from paying attention to nature. It is largely about respect. The morale of the colony is paramount, and any creature needs to be comfortable. You must get the bees out of drafts and make sure they have enough water. Drought comes from the nature of the subsoil. Some land needs rain every ten days, or the plants become stressed and, in turn, the bees suffer. In nature, nothing is fixed— the weather, plants, geography—all have their variations. When Willie was learning from his father, they had a fixed routine of visiting the hives. Sometimes the bees would just not be in the mood and the situation would turn nasty. Now he goes more with what the bees need and want, rather than dictating terms.

After spending some time with Willie, I began to understand what is at the heart of his philosophy. Many other beekeepers have gone down a more scientific route, which involves breeding up or bringing in new strains to promote positive genetic traits in their stock. After making a good living from beekeeping for thirty-five years, Willie has learned the hardest lesson of all: to do less, not more. He lets the bees adapt to their environment the best they can. He visits them occasionally, he checks them for disease, he makes sure the hives are in a good position, and he collects the honey. Only after years of practice, observation, and knowledge has he reached this conclusion: that the best you can do is to do as little as possible.

∾

FIVE ROOMS of the honey farm are given over to an exhibition that covers every surface with information on the honeybee. As well as displays on the biology and ecology of bees, there are bee-related quotes ("If you want to gather honey, don't kick over the beehive"—Abraham Lincoln); bee facts about history; and snippets on bees and beliefs. The insect is featured in the Bible, with its land of milk and honey, in the Koran, and in Native American and Hindu creation myths—in the texts of almost every major religion and literature in the world.

At the end of the first room I discovered an observation hive. Aristocrats of the Enlightenment had a craze for these glass hives as they began to look at the world more closely, with an eye cleared by reason. The hive was about 4 feet wide by 3 feet high, containing a flat slab of comb. The bees went in and out through a tube in the wall. I stood before it, mesmerized. There was a faint sound, like a nonstop train going through an endless tunnel a mile below my feet. This animal engine chuffed away in a ceaseless steam of pure energy. I put my ear to the glass and both felt and heard the whirr of life: thousands and thousands of lives wound up like watches, ticking away in collective survival. I blurred my eyes. The bees formed an almost solid material, quietly, steadily seething. It was unlike anything I had seen before. Not repulsive, like the pulsing of maggots on meat. Not a crawling, or scurrying, or wriggling. It had a gentle, purposeful, cohesive movement, impressive and unstoppable in its numbers, like a crowd gathering at a large sports stadium, or a workforce funneling into a factory gate.

I thought of Willie Robson and how he and the bees lived with their surroundings; of the intimate and ultimately curious relationship that holds bees, plants, and humans in so many different, delicate webs around the world, and how these webs had existed for millennia. The products of the honeybee have always been useful to man, but beyond, far beyond, their material use, bees have always fascinated us, right from the Stone Age cave dwellers who drew images of wild honey hunts on their walls, the primeval lines flowing with a life that is simple in the way that water is simple, or blood. The life of the hive has provided metaphors, as we compared this insect society to our own, and it has always provided mysteries, whether we tried to understand them or fed them into our superstitions.

Civilizations rose and fell; the bee flew on regardless. How man has seen bees reflects back an image of each age. I looked at the bees in the observation hive, both individually and collectively, trying to understand this essentially strange sight. Thousands of human voices—poets, scientists, saints, hedonists—joined the hum: bees must be the most discussed creature on the planet, after man. A buzz collected, over centuries. I did not know what it would tell me, but I knew, then, that I wanted to write this book.

CHAPTER ONE

IN THE BEGINNING: EVOLUTION

Honey. It starts in the spring. With the brightening air comes a quickening of the world. All over the planet, plants plug into the energy of daylight. Systems are switched on; leaves feed on light; sap circulates. As spring spreads to summer, flowers in uncountable quantities open out. Within the plants lie small, secret pockets of nectaries, and within these glands swell droplets of sweet liquid. This sugary substance is a symbol of all that is desirable in nature: nectar.

The female worker honeybee hovers, lands, and bends into the center of a flower, head down for a feed. She sucks up the nectar, then she's off to another one—accelerating so fast your eyes are left behind. She collects the nectar in a transparent, pear-shaped bag called a honey sack that lies at the front of her abdomen as part of the gut. When this sack bulges full, she flies back to her colony.

In the dark of the hive, this forager bee passes the nectar on to the house bees. The nectar will be passed from bee to bee, becoming progressively more concentrated as it goes. The bees push the nectar into flat drops on the underside of the proboscis, and exposure to air helps evaporate some of the liquid. A drop will be pumped in and out many times, each time becoming a little less liquid. Sucked and pumped, sucked and pumped, sucked and pumped, the nectar concentrates down to 40 percent of its original moisture, and then small droplets are deposited onto the floor of the wax comb where the warm air in the hive evaporates it yet further. As foraging bees bring back the sweet flow, thousands more beat their wings to create a through-draft; the colony is a mass of wings working together to fan off the moisture with warm air. When the liquid reduces right down, and each cell's watery gleam has thickened to a sticky bead, the bees top the full cell with a wax cap. Sealed and stored, the honey is now ready until needed, rather as you might keep a pot upon a cupboard shelf.

But wait. This image of the well-ordered household with its well-stocked larder is too tame, too neat, too cozy. It is far less strange and extraordinary than the truth: the miraculous has been domesticated. For what is honey, once you take it off the shelf and trace back to where it comes from? Each place, each plant produces a different honey. Honeys have tastes, colors, and consistencies according to their nectar sources. Some honeys come largely from a single flower—monofloral honeys—while multifloral honeys gather the nectar of many plants from places such as meadows and mountainsides. From the tough scented carpet of thyme and marjoram on the slopes of Greek islands flows a nectar that becomes a honey that was once offered to gods. The tree-of-heaven's honey tastes faintly of muscat grapes. Bees fly between orange blossoms and the splayed white flowers of the coffee plant, fusing their flavors as they go. They fly to milkweed, thistles, and goldenrod; to dandelions and tulip trees, to acacias and rock roses. A slightly salty, snow-white honey comes from the pohutukawa, the Christmas tree of the blazing, antipodean midsummer, that flowers flame red around December. Frothy white blossoms on apple trees produce orchard honey. The violet, snaky stalks of viper's bugloss make a clear gold honey. Italian chestnut trees spread a dark fragrance; mango honey is truly fruity, and the *aguinaldo blanco* of Central America yields a water-white honey said to be one of the clearest in the world. Fields

of lavender, of beans, of oilseed rape; suburban gardens full of flowers nodding with bees; Californian desert and Himalayan cliff; the
bone-dry Kalahari and the looping, raveling rain forest; the fairy-
tale dark-depths of woods in Central Europe with their resinous
honeydew; carob plantations and Sicilian lemon groves; rambling
British blackberries and the many different kinds of eucalyptus in
Australia, which flow unpredictably, perhaps every two, eight, or
twelve years; the rich, dark resonating brown of rosemary honey;
the slightly minty honey from the linden trees on the Lower East
Side of New York City: all these plants, all these places, stream with
nectar in large gouts or pinprick stars; all come through bees to
make honey.

It is closer to the truth to say that bees perform an act of alchemy.
Honey is nothing less than concentrated nectar; and a pot of good
honey is the essence of its surroundings, a sweet, fragrant river
from a million tributaries, carried across the air and flowing gold
into the pot through the transforming power of the bee.

THE HONEYBEE'S STORY must be traced back through an
incomparably vast stretch of time, through clues strewn in the great
evolutionary flow. The search feels like a detective story. Where to
begin? The canvas is unimaginably large. Life—the chemical
change that sparked inanimate matter to reproducing molecules—
probably began about four billion years ago. (Humans, of the sort
we would recognize, have probably existed for about 1.5 to 2 million years, to give some idea of how insignificant we are in terms of
time.) The next stirrings of existence began in the water that surrounded both plants and animals, bringing them food and oxygen
and supporting their bodies. Life moved from seawater to fresh-
water, creeping further toward land and then colonizing its
swampy margins. A coating of primitive plants moved across the

earth. It was insects that evolved to feed on these plants, and their remains have been found in fossilized swamps and remnants of the earliest forests.

Early insects were wingless; then, as the plants grew, they developed wings that could more easily reach the new heights. Bees, like ants and wasps, are part of the Hymenoptera, or "membrane wing," order, with two sets of filmy wings hooked together to cause less turbulence and drag in flight. The wings are stretched over a sparse network of veins that provide their support structure, like the frame for a kite's flexible fabric.

Evolution is the blind shuffle of DNA, filtered by success of reproduction. Insects have succeeded by being the ultimate niche operators of the animal kingdom, able to work in any environment, from Arctic wastes to mountaintops to suburban gardens to deserts. One reason they can do this is that their exoskeletons can adapt relatively easily, the animal's outside altering without the insect's inside having to change. The exoskeleton adapted into different kinds of wings; it turned into the needling legs of the spinning spider and the musical saws of the jumping grasshopper; it became the warning spots of the ladybug and the aggressive stripes of the wasp and the bee; it became the battling claws of the stag beetle and the stabbing jab of the mosquito. The insect has an external kit that tools it up for many different circumstances and its evolutionary success is proved by the numbers. There are a million insects for each human on the planet, and they make up around half of all named species.

How did some insects become bees? The first clue is their intimate connection with flowering plants, or angiosperms, which arrived on earth during the geological era known as the Cretaceous, between 140 and 60 million years ago. Primitive plants spread their seed by wind, casting their pollen into the world in profligate quantities. Then some plants began to make smaller quantities of pollen

than their predecessors and invested more energy, instead, in entic-
ing creatures such as insects to visit. It was a smart move. Insects
evolved to feed on the protein-rich pollen, the tiny grains that are
the sex-dust of male reproduction. When pollen attaches to an
insect's body, it can be transferred to other plants and—bingo!—
pollination occurs. When you want to attract lovers, it pays to dress
up. Flashy, colorful, sweet-smelling flowers evolved, appealing to
animals, and particularly insects. Nectar, the base material of honey,
is part of the flower's tactics of attraction, along with petals, pollen,
scent, shape, and color. Honey, then, is an elixir of sex.

THAT FLOWERS EVOLVED at the same time as many of the
insects must be no coincidence. Bees and blooms are so twisted
together by the twin necessities of existence, of reproduction and
food, that their development must have been interdependent. The
chronology of this is not entirely certain, however. The clues of
paleontology can literally be writ in stone, yet they are still random
clues to life, and petals and insects preserve far less well than
dinosaur bones.

The oldest known bee fossil was found in New Jersey. This single female insect is entombed in the hard, orange glow of amber.
She was, poor scrap, trapped by sticky coniferous tree resin. She
was also captured for posterity. The resin turned to a light, transparent fossil and the bee was held forever, legs stretched out, almost
flailing, as though she is either tumbling through some other-
worldly medium, or about to land on a plant that produced the
pollen of eighty million years ago. The bee is caught in a fossilized
freeze-frame, the durability of the rock starkly framing the delicacy
of the fragment of life within. She dates from the late Cretaceous
and was already well evolved, evidence pointing toward the fact
that bees had been around at least as long as flowers.

Then, in 1994, a discovery was made that could push back the date of the evolution of bees even further. It raised the idea that they could have been on the planet perhaps even *longer* than flowers.

The Petrified Forest National Park in eastern Arizona is a time capsule of stone logs gradually being uncovered by erosion and explorations. The 100,000 acres once contained the Black Forest of ancient conifers that thrived in the semitropical world of the Triassic period, more than 200 million years ago. Then volcanic eruptions sent a huge flood that flattened the trees like skittles and buried them deep underground, devoid of oxygen. Over time, the wood started to mineralize. In some cases, iron oxides in the wood turned the trees into a startling range of colors such as ruby brown and lichen orange; in other cases, they stayed as black as the forest's name.

Time passed. The landmass of planet earth that had been a supercontinent split into a northern half, Laurasia, which later became North America and Eurasia, and a southern half, which became South America, Africa, Australia, peninsular India, and Antarctica. Humans arrived. Humans evolved. Humans became curious. Humans became acquisitive.

By the nineteenth century, the fossilized forest had gained a certain celebrity. On the orders of the Civil War commander General Sherman two petrified tree trunks were carted off to the Smithsonian's National Museum of Natural History, where they remain today. Amateurs and professionals also came to the forest, picking up souvenirs and booty, from shards to logs. The petrified fragments were turned into clock bases, jewelry, and luxurious trinkets sold at Tiffany. In 1962, President John F. Kennedy made the Petrified Forest a national park, affording it some protection.

At the end of the twentieth century, teams studying ancient ecosystems and climates tracked through the park, trying to gather clues about the forest's original existence. Among the most interesting finds, of a group led by Dr. Tim Demko, were approximately one hundred insects' nests. The inch-long flask-shaped cells were clustered together, and the entrance was probably through open knot holes in the wood. The formation of the cells and details of their constructions led the scientists to believe they were built by ancient ancestors of today's bees. Elsewhere on the site, they also later found nests closely resembling those of the modern sweat bees (Halictidae), so called for their attraction to perspiration. Chemical analysis of the Petrified Forest nests showed that the cells contained some of the organic compounds found in beeswax.

The early date of the forest could be significant. If these were, indeed, bees' nests—and the evidence certainly pointed to this, though some say you would need to find bee bodies to be certain—it would mean bees existed 207 to 220 million years ago, at least 120 million years or so before the oldest previously known bee fossil. Beyond this, the nests are older than the earliest known flower fossils. Could it be that bees existed, in some form, for ages *before* flowers?

It depends, partly, on how you define a bee. Evolution is, after all, a continuum, and these could be bee ancestors rather than bees themselves. It also depends upon when flowers first evolved, and fragile plants leave an elusive fossil trail. Charles Darwin called the origin of flowering plants "an abominable mystery," and it remains, for all the theories, ultimately mysterious.

What seems more certain is that bees probably evolved from a descendant of today's carnivorous hunting wasp. The Russian entomologist Professor S. I. Malyshev posited a theory about how this happened, and this leads to part of what makes a bee a bee: its diet. Bees are unusual among insects because the developing young have the same diet as the adult; both survive exclusively on plants. The hunting wasps still feed their grubs on protein-rich aphids that they kill with their jaws. They also eat, as bees do, the honeydew exuded

from plant-sucking aphids. Malyshev argued that these early carnivorous wasps, in the process of killing their prey to feed their
young, would taste the sweetness in the aphid's body that they also
found in honeydew. It would have been a short evolutionary step
for the insects to feed entirely on plants, in their larval as well as
their adult diet.

We do not know for certain if and how the bee first evolved from
a carnivorous hunting wasp. We do not know the earliest date of the
bee or the bloom. What we have are theories looped onto fragments.
It makes the evolutionary detective story no less intriguing.

To get back to a concrete fact—one you can eat—Malyshev's
speculations connect to another pot in my kitchen cupboard. Honeydew honey is a delectable curiosity. Strong, to the point of almost
being savory, it is not made from the nectar of flowers at all. Rather,
bees collect honeydew from the aphids in forests, just as the hunting
wasp did all that time ago. This honey is therefore a sticky substance
made from fluid ingested by two kinds of insect. But, as I spread the
darkly delicious ooze on my toast, I prefer to think of honeydew as
a possible clue to the evolution of the honeybee.

~

THE COMPLEXITY of the relationships between bees and flowers shows that they certainly coevolved to a great extent. Many
flowers have highly specific structures to attract bees and other
insects. The colors and shapes of flowers that so appeal to the
human eye act on the bee, though in a slightly different way. The
velvety red of a rose, for example, is wasted on a bee, in terms of
color, because their eyes cannot distinguish differences at that end
of the light spectrum. Toward the yellow, green, blue end, however,
a bee's sight is acute, which may explain why such plants as thyme,
rosemary, and lavender, with their blues and purples, are such
famous sources of honey.

At the far end of the light spectrum, bees can see the ultraviolet light that is invisible to the human eye. Seen through a bee's eye, the apparently uniform yellow of the petals of the evening primrose will reveal markings that guide the bee into the center of the flower. Other flowers often have a different color at the center—such as the yellow in the center of a rock rose—which also leads the insects toward the nectaries and the pollen-bearing stigmas.

Smell is another way that flowers attract bees and help them to remember to return. Scent and sight clearly work well, because for approximately twelve days, the honeybee retains the knowledge that a particular flower yields good nectar and pollen. This helps promote what is known as constancy, the return visits that benefit both bee and flower. The flower is more likely to get pollinated if the bee is going from one to another of the same species; and the bee is able to suck up as much of a good source of nectar as possible. The bee has a daily diary of appointments since flowers tend to produce their nectar at set times. Bees develop an ongoing schedule that changes as different flowers open and deliver nectar. Nine in the morning is good for dandelions, while marjoram flows at lunchtime and viper's bugloss gets going at 3 p.m.

Bees' bodies have evolved in a way that makes them able to work flowers. The basic division of an insect's body—into head, thorax, and abdomen—led to the name "insect" or "cut into." Bees have, in addition, what you could call a wasp-waist (in insects as in corseted Edwardian ladies); that is, two segments of their abdomen narrow to an adjoining point. This means their bodies articulate so they can poke their front right into a plant. Flowers with downward swinging bell-like flowers, or flowers with oddly shaped or narrow tubelike structures, are easy pickings for a bee, whilst inaccessible to less flexible creatures.

Some connections between bees and blooms are extraordinary in their specificity. The most celebrated is that between the euglos-

sine, or orchid, bees and the bucket orchid of the *Coryanthes* genus
in South America. Euglossine bees are among the most beautiful
bees on the planet. Their metallic abdomens are like shards of
enamel in hues of blues and greens, bronzes, and golds, darting
through the air in rapid flight. For their part, orchids are famous
for an enormous variety of forms, and this is connected to their various, unorthodox methods of reproduction. Few sexual relationships are as strange and difficult as this.

The flower of the bucket orchid operates like an assault course.
Its structure, odd enough to be beautiful, looks more like a diges-
tive tract than the stereotypical idea of a flower. At its base is a deep,
bucketlike form, into which drips fluid from two taplike protuber-
ances above. The flower emits a scent that draws the euglossine
males. The bees land on the rim of the bucket. The flower's whole
design then encourages a bee to fall in. The steep, smooth sides of
the bucket are impossible to climb. The bee has only one route out:
a spout situated halfway up the flower. In a bid to escape drown-
ing, the bee pushes its way through this narrow exit. In the process,
he brushes past a packet of pollen hooked onto the top of this
exit corridor, which should, fingers crossed, attach itself to his
abdomen. If this bee then flies to another bucket orchid and the
same incident occurs (will these bees ever learn?), he has to scram-
ble free again, pushing through the escape passage and in the
process losing his pollen packet so that pollination occurs. No wonder bucket orchids are, even for orchids, comparatively rare.

FROM THE EARLY ORIGINS of the hunting wasp, there evolved
many ways of being a bee. There are, today, at least 22,000 named
species. Because nectar, like nature in general, flows well in
warmth, the greatest number are found in the humid tropics.
Brazil, for example, has at least four thousand of them. But bees

also thrive in hot, dry places such as deserts and can survive in such apparently inhospitable places as the Himalayas—certainly as high as 14,760 feet—and Arctic tundra. The larger the area, the greater diversity: an island such as Great Britain has 260 kinds of bees, France has 800, and the African continent and North America each have 4,000.

This superfamily of bees ranges in size from the smallest bee in the world, the *Perdita minima*, dozens of which could fit on a single antenna of the largest bee, to the big, black leaf-cutter bee *Chalicodoma pluto*, with a body one and a half to two inches long. Some groups have been given names mirroring human activities. Miner bees, for example, dig deep into the ground—one Brazilian species as far as 16 feet down—making mininests at the ends of tunnels dug off a central shaft. They dig like dogs, their legs throwing up dirt behind them, and observant gardeners, noticing the small mounds of earth on their lawn, mark this as a sign of spring. Mason bees mix dust with saliva to form a cement to construct cells for pollen and honey. Carpenter bees use their strong jaws to bore and cut their way into wood and hollow stems to make their homes, and leaf-cutter bees snip semicircles out of leaves, puzzling gardeners with these large, neat munch marks, and fly with the leaf grasped by all three pairs of legs, taking it to line their nests, which are found within such prefab spaces as beetle tunnels, plant stems, and even animal skulls.

Bees have reached different degrees of communal living. Bumblebees in temperate climates, for example, form colonies in the summer. Most of the bees die off over winter, leaving the queen alone. She finds a ready-made hole, such as a mouse's abandoned nest, and spends the winter there with a single cell of honey. When the weather warms and the nectar starts to flow, she eats some honey from the pot and leaves her nest to start the year and a new colony. The sight and sound of bumblebees is a clear, early sign of the shifting up of the year's gears toward longer, warmer days ahead.

Of the many ways of being a bee, the honeybee, as its name suggests, is distinguished by its high degree of communal, or social, behavior, which means it has become extraordinarily efficient at producing and storing honey.

The nine species of the *Apis*, or honeybee, genus all have highly social colonies and nests of hexagonal, wax cells. By collecting honey so effectively, they can survive a drop in the nectar flow and other forms of adversity. *Apis florea*, the dwarf honeybee, is about ¼ inch in length, whilst *Apis dorsata*, the giant rock bee, can be over ¾ inch. Both are indigenous to southeast Asia, where they build single, large combs in the open. The comb looks like stiff swags of curtain hanging off objects such as tree branches. *Apis dorsata*, with its shaggy, long-haired coat, can survive cooler heights and is the bee of the Himalayas, where it builds its comb on cliffs—a comb that is big enough for the honey hunters of Nepal to roll up and take back to the village as a great prize.

There evolved two more branches of the *Apis* genus, which seem to have existed for just a tenth of the time of the open-nesters. Both kinds of bees began to build parallel combs in cavities such as hollow trees. Furthermore, and most significantly, both evolved the ability to form an inert cluster. This meant the colony could survive colder winters because this cohesive, clinging ball of insects regulated its own temperature, loosening when the colony needed to lose heat and pulling together to conserve it. The colony no longer died off in cold weather, which meant they could move beyond tropical zones into a far, far wider geographical area. They also had sufficient numbers to start collecting nectar as soon as it started to flow again. Building a nest in the dark was an ability that was later to lend itself to living in hives.

Of these two kinds of bee, *Apis cerana*, the eastern honeybee, is native in Asia. Its colonies tend to be composed of six to seven thou-sand bees. The other bee spread through Africa, Europe, the Middle East, and western Asia. It can build and sustain colonies of as many as 100,000 or more and has a prolific and consistent rate of honey production. Being suited to hives and producing such quan-tities of honey, this bee was all set to play a role in the life of man. It was to become the superbee of planet earth: *Apis mellifera*, the most successful bee of all time, had arrived.

APIS MELLIFERA, the most studied creature on the planet after man, is a summit of sophisticated engineering; an evolutionary tri-umph of form and function in a thousand details. This tiny crea-ture's achievements tower above the flights of architecture and efficacy of man-made machinery. It has occupied the minds of sci-entists, writers, musicians, and philosophers around the globe.

The fluffiness of bees—even a bee's eyes are hairy—is seen by humans as part of a cartoon "cuteness," but the hairs that cover the honeybee's body serve many purposes. They help create an electro-magnetic charge that draws pollen grains through the air, leaping toward the bee's body, where they are caught in the mesh of strands. The bristles growing between the 6,900 hexagonal plates that make up the compound lens of the bee's eye help it to gauge wind direc-tion and flight speed. Hairs also cover the six pairs of legs. Forget bees' knees; bees' legs, on the other hand, make humans' look pedestrian. Each one ends with a claw and suction pads that enable the bee to move horizontally and vertically, to land on a petal and cling to it with just a single leg, or to hold onto other bees to form the cluster of a swarm.

The legs are hairy, and each and every bristle has its business. The forelegs have a small, hair-lined notch through which the bee cleans its antennae, keeping their sensitivity clean and bright. The

A worker bee returns to the hive with willow pollen packed into her pollen baskets.

lower, outer part of each hind leg has a concave scoop, into which the bee packs its pollen, the grains moistened with sticky, honeyed saliva so as to form small clumps. Stiff bristles of hair help anchor the collected packet of pollen, and the honeybee has evolved a sweeper system of moving pollen grains down its body to collect in these pollen baskets, which you can sometimes see when it rubs its legs together in flight. You can certainly see the pollen clumps with the naked eye, if you look carefully at the hind legs of a bee as it pauses on a flower, or even as it flies around the garden.

The color of the pollen bundles depends upon the flowers the bees have visited and is one way beekeepers can tell where their insects have been feeding. A beekeeper's pollen chart looks like a paint catalog with some surprising matches of colors and titles. Snowdrop pollen is the color of a free-range chicken's egg yolk, and red dead-nettle a sultry, vampish red; but asparagus is the orange of a 1970s plastic chair; raspberry, for some reason, is gray; gorse is mouse brown; and oriental poppy is dark blue. Between these are many shades of green, gray, orange, red, yellow, and brown.

The honeybee's wings are a transparent film, as strong as they are light. Although they look like just a single pair, each bee has two pairs, the forewings linked to the back wings by a row of minuscule hooks that link the two together to make a hard-wearing yet flexible whole. The muscles in the insect's thorax enable the wings to beat at an astonishing two hundred times per second, and it is this rapid, energetic movement that makes the bee's buzz. This tremendous energy means the thorax could overheat, but the excess heat passes to the head, where it is dissipated and evaporated in droplets of semi-concentrated nectar, as if the bee were sweating a watery honey.

Honeybees are proverbially busy; it is even more impressive that they don't waste energy. Even a casual look at an observation hive shows a mass of activity, but just as much waiting around. Up to two-thirds of a bee's life is spent wandering around the hive doing pretty much nothing. You wouldn't call it relaxing, exactly, just not using up energy on the unnecessary.

Where there's a good load of nectar and pollen to be found, they go. The average speed of a bee in flight is 15 miles per hour. Its fuel is honey, with an average load lasting up to 37 miles. A gallon of honey petrol could take a bee seven million miles. But the bee is careful not to overload. It calculates exactly how much honey it needs to take into its honey sack in order to fly to and from the forage flowers without weighing itself down needlessly.

A beehive with its honeycomb and its bees is an embodiment of energy: energy recycled endlessly as the honey is eaten and used in the form of flight and every other facet of being alive. As interesting and impressive as bees are individually, it is as a collective force that they most fascinate. Each bee may be a tiny fraction of the hive, but each one plays its part. The collective life of the hive enables it not just to thrive but to grow, with new colonies of bees breaking off, or swarming, to create new colonies.

We tend to put a hierarchy on the three types of bee found in a hive. While there may be up to 200,000 female workers, though more commonly 50,000, there might be, at most, a few hundred drones, or male bees, and just one single queen. Her singularity and reproductive function—she is the egg layer for the colony—crown her as the most important bee. When a nest becomes overcrowded, the colony prepares to swarm. This involves the old queen flying off with some of the workers in a great swirl of bees. Before she goes, more queens will be "started," one of which will lead the bees remaining in the colony; others may take off with further groups or casts, depending on the needs of the colony. The queen cells, in which the new queens grow, look a bit like thick, waxy, inch-long peanut shells and hang vertically from the comb. When no more queens are needed, the next queen to hatch will destroy any that follow her. An eerie, keening noise can be heard outside the hive of the queens calling out, about to be born. They should keep quiet. The hatched queen will wait for the others to emerge and the ensuing fight is a sting-to-the-death. The queen's sting, unlike the worker's, can be retracted and reused, ruthlessly.

The new queen's next major task is to mate. She leaves the dark security of the hive for her virgin flight, to be pursued by the male bees, or drones. They mate in midair and she returns to the colony, storing the sperm to use during her lifetime. From then on, the queen's business is mainly to lay eggs—as many as two thousand in a day. She may be able to lay a million or more eggs altogether. After she has run through her supply of sperm, her end is near, and

Queen   Drone   Worker

The three kinds of honeybee: queen, drone, and worker.

she begins to look a bit threadbare. At this point, the colony starts a new queen by feeding eggs with a high concentration of royal jelly, a protein-rich food derived from eating pollen.

The queen's life shows how the power and abilities of the individual bee extend only to the greater good of the colony. Nowhere is this more poignantly obvious than in the life of the drone. P. G. Wodehouse named his fictional Drones Club for idle young men with good reason. For much of his life, a drone could be portrayed, anthropomorphically, as a lazy so-and-so. He flies out on sunny days, around midday, but not to collect food. He doesn't even bother to get food for himself in the hive but is waited on by the workers. The drones even excrete in the hive and the workers clean up after them.

The whole point of the life of a drone takes place when the new queen is ready for her flight. Because, in evolutionary terms, mating is the crucial part of his life, a drone's anatomy is geared toward this single event. Blunt, square, and altogether stockier than the worker bees, he cuts a masculine figure with his large eyes like the aviator shades on a strutting rock star. These eyes are important for following the queen. His huge eyes, and his powerful antennae, ten times more sensitive than a worker's, are not there to help him find flowers but for this mating-on-the-wing.

This, then, is his mission in life: to catch up with the queen and mate. When the front-flying drones achieve this feat, they die. From this point on, the rest of the drones are marked men. Charmed as their life may seem, they are pointless. Although they may spend the rest of the summer hanging out in the colony, or making their sunny, midday flights, when the colony is closing down for the winter, or under stress of any sort, the workers simply stop giving the drones food. They even bar them from coming back into the hive when they return from their flights. A beekeeper may see the pathetic sight of dozens of drones shivering outside the hive on an autumnal day. Denied food and shelter, shortly they will die. Next spring, a new lot of drones will be born, to grow, live, fly, feed, mate, and die.

Crucial though the lives of the queen and drones may be, the *real* power behind the hive is not the bees that do the mating, but the tens of thousands of female workers. To call them workers is to see them as drudging laborers who lack the glamour of the queen and drones. Yet the lifetime of a worker bee contains a complexity and progression that makes the other two kinds of bees look almost mundane.

The worker bee is the ultimate multitasker. From the moment she is born to the moment she dies, she performs a series of widely differing tasks: cleaning cells, tending and feeding the larvae and pupae, building and repairing the honeycomb and nest in general, receiving nectar from foraging bees and further processing it into honey, receiving and packing pollen into the cells, ventilating the hive to keep it at the right temperature by flapping her wings, guarding the entrance to the hive, and then, about halfway through her life, going out to forage flowers, bringing back nectar and pollen, reporting back good finds to the hive, making honey, eating some honey herself, and going out again to forage some more. A worker bee in the summer may live for six weeks before dying, eventually, of exhaustion. In her life, she may have collected enough nectar to make just ¼ ounce of honey, less than half a teaspoonful.

How the worker bee ticks off all her tasks is just one of the extraordinary achievements of the honeybee. The amazing complexities of these bees and their colony are the subject of many centuries of discovery. How do these thousands of insects communicate? How do they, for example, know when to leave the hive in a swarm? How do they find the flowers that will produce honey and bring others to the feast? How do they build the wax comb, which can hold the largest amount of honey in the lightest and most economically designed structure possible?

The honeybee evolved, flew around, and reproduced without name or number. As the Ice Ages locked up water to create land bridges, bees spread farther. In the north, they spread to the tip of what is now called Sweden; to the west, to Ireland; to the east, beyond the Ural mountains, and to the south, to the tip of Africa. In each place, *Apis mellifera* evolved slightly differently to suit the area, their limits at first marked by geographical boundaries of seas, mountains, and deserts. Then another creature appeared that was to change everything. Enter, *Homo sapiens*.

CHAPTER TWO

WILD HONEY

At first, humans found food using the same methods as our fellow animals: hunting and gathering. Berry by berry, kill by kill, we pieced together the nutrients necessary for survival. We fed like animals, we fed on animals, we scavenged from the kills of other animals. Doubtless we learned techniques from them, too.

Honey-eating is an appetite we share with other mammals, particularly primates. Mountain gorillas on the border between Rwanda, Zaire, and Uganda hunt for wild honey, as do monkeys and baboons. Chimpanzees, our closest cousins, with whom we share 98 percent of our DNA, are adept honey hunters, cooperating to raid bees' nests of wild honeycomb and sharing the spoils. Some of their honey habits sound remarkably human. In her observations on chimpanzees in Tanzania, the primatologist Dr. Jane Goodall describes a male chimp using a stick to enlarge a hole in the opening to an underground nest. He pulled out the comb and shared it with his mother. Fifteen minutes later, his baby sister came back and repeatedly put her hand in the dripping honey, licking her hairy fingers like a child who has discovered the family honey pot.

Bears are greedy plunderers of honeycomb. From fictional Baloo to Winnie the Pooh, from brown bears to the sloth bears and sun bears known as honey bears, they relish the sweet wealth of energy this delicious food provides, liking both the honey and the protein-rich developing brood. Particular bears are such expert honeycomb raiders that they may be marked out and shot by humans protecting their hives.

Of all the tastes known to man, sweetness has the highest status. Bitterness means bile and seething rancor, as well as coffee; saltiness can swell into a gagging choke; sourness can mean putrification rather than the bright edge of lemon. Sweetness, alone, is heaped with connotations of wealth and happiness. You might eat one too many chocolates and feel sick, but that, after all, is a problem of luxuriance.

This high regard for sweetness is based on sound nutrition. For early man, sweetness in fruit proved it was ripe and ready to eat. Sweetness releases sugars quickly into our bloodstream for instant energy, and honey, with its easily digested fructose and glucose, not only does this fast but also provides an especially dense source of calories. A stash of golden honey would have been a bonanza to early hunter-gatherer humans, compared with the patient grubbing and picking together of leaves, fruits, and roots or the difficult hunting down of wild animals.

Our attraction to sugar is something we are born with. Sugar occurs in nature in plants and mammals' milk, and experiments have shown how newborn palates turn to sweetness regardless of experience, suggesting that it is an innate need, soon supplied by breast milk. Of all the sources of sugars readily available from nature, the honey derived from nectar tastes especially sweet, making it a particularly nurturing form of nourishment.

Food is not only a matter of nutrients, or perhaps we would today be living the space-age fantasy of pill meals. Even for Stone Age man, there were easier sources of calories than robbing wild bees. The way we love honey must be connected to our positive feelings about sweetness in general, perhaps connected to our earliest nurturing. "Good food," wrote the anthropologist Claude

Lévi-Strauss, "must be good to think about before it becomes good to eat."

We know that honey was revered by early man, was woven into our highest thoughts and desires, because, along with the other animals most prized in hunting and mythology, the honeybee features widely in the mesmerizing, earliest surviving marks we made about the world: ancient cave and rock drawings.

IN THE ICE AGES, when European humans retreated to live in southern France and the Iberian peninsula, their hunting quarry was not just the hairy-coated mammoth and bison. A limestone cave in Spain contains the first painting discovered in Europe to be recognizably of honey-hunting. Found in 1924 in a rock shelter at Bicorp, inland from Valencia, the image depicts a set of long, swinging ropes, let down over a rock face—like the picture itself is, in a literal sense. Halfway up, one climber clings on, as the ropes billow out slightly under his weight and momentum. Right at the top, another figure reaches into the nest of wild honeycomb; his other hand holds a bag with a handle, such as those made of the stomach of an antelope's hide, used by later hunter-gatherers to enclose the honey and keep the bees off. Around a dozen bees funnel in toward this thief, wings spread in a buzz of activity as they come from different directions, moving freely through the air, as is the pattern of disturbed bees rather than the more directed mass of a swarm. The humans, small streaks on the long ladder, have far to fall, while the insects are as big as, or bigger than, the hunters' heads, adding to the hunters' vulnerability. The image expresses the scale and danger of the endeavor: the perilous, swaying ropes; the people dwarfed by the climb and the insects; the threatening focus of the bees' flight; and the ripe, heavy fruit of the prized honeycomb at the top.

Honey-hunting rock art, from Bicorp, near Valencia, Spain.

The power of this remarkable image is easy to feel, and harder to analyze. Although painted somewhere between two thousand and eight thousand years ago, the prehistoric humans and their honey-hunting appear today as if the time between them and us had dissolved. After the initial shock of recognition, what the contemporary mind reaches out for, like a sound just beyond hearing, is the meaning of these images to their makers.

Cave and rock drawings often depict animals, and this has led to the suggestion that the images were some sort of magic charm connected to improving the chances while hunting. But this is probably only part of the truth. Animals are not the only images in such caves. In some places there are poignant pictures of human hands, left like ghosts on the walls by the artists blowing paint around them. There are other images, too, particularly geometric patterns such as dots and lines, which do not fall into this literal reading of cave art as portraits of prey. Many of these paintings were made in inaccessible places, deep and dark within hillsides. For all these reasons, it is probable they were not homes, but more likely places of special significance for ceremonies of some sort.

Another significant Spanish cave drawing site is on a farm called Altamira, or "high lookout," near Santander in Castile, a couple of miles from the north coast of Spain. It was found in 1868 when a hunter's dog ran into a cave and was pursued by its master. The farm's owner, Don Marcellino de Sautuola, an amateur archaeologist, saw little apart from bones when he first went to look at the site. When his interest in the Ice Ages was later aroused, he went back with a team to explore the caves more thoroughly. Sautuola's explorers crawled and pushed deeper and deeper into the rock, to discover that the caves extended to about a thousand feet of galleries and side chambers. One day, Sautuola's young daughter Maria joined in the search. Whilst her father had been forced to crawl through one of the low chambers, Maria could easily walk in and look up at the ceiling. The magical excitement of seeing paintings deep within caves, by the flickering light of a candle, must have felt even to a young child like being transported back to the age when they were first made and seen by firelight. Twenty-four bison thundered across the rock. Around them were horses, boars, deer, and a wolf. It took twenty-three years for the drawings to be accepted as authentically Ice Age because academics just could not believe that such powerful images could be made by early man.

There were further discoveries to be made about the images in the Altamira caves, and it is these that relate to bees. We are familiar with the animals in cave paintings because we recognize the subjects, and these are the pictures that tend to be printed in books and on tourist postcards. What are less well known, and far harder to understand, are the geometric patterns. One such image lies in the caves at Altamira. It depicts a number of concentric semicircles, almost like a child's drawing of a rainbow turned upside down. Strangely, this exact image is found in many cave-drawing sites in southern Africa. Experts believe this image may well be connected to wild-honey-hunting and that it can give us vital and intriguing clues into the meaning of bees and honey to Stone Age humans.

THE GREATEST CONCENTRATION of honey-related rock drawings in the world is in southern Africa. The continent, as the birthplace of mankind, has more cave art than any other part of the world. Dating it is an uncertain business, but there is some evidence of such art a staggering 800,000 years ago. There have been art-related tools discovered from 40,000 years ago, and paintings and carvings that have been dated back to 15,000 to 20,000 years ago.

Within the four thousand rock-art sites in Zimbabwe and South Africa, there are twelve images specifically of honey-hunting, and two hundred sites in Zimbabwe alone that may well allude to honey. These images show ladders, bees in flight, even a hunter holding flames to smoke the bees out of their nest, as is done today. More abstractly, there are nested semicircles, such as have been found in Spain, some of them engraved into granite, and another, often repeated image of long, plump, adjacent bulbous shapes, almost like a collection of long grubs. Some rock-drawing experts call these shapes formlings; their shape is instantly recognizable as something organic, but what exactly it represents is not at all clear at first sight.

However, honey experts have come up with an intriguing interpretation of their meaning. They turned away from the rock face and back to the living world to examine wild honeycomb, which looks today exactly as it would have in prehistoric times. From the front, as if you had just hacked vertically into a hollow tree trunk and removed the front like a panel, wild honeycomb looks like a series of parallel, semicircular curves, the largest in the center, with the combs getting progressively smaller toward the outside. In other words, they closely resemble the nests of curves found on the rocks. If you look at the honeycomb from underneath, the shapes are again familiar: they bear a remarkable resemblance to the organic formlings. Like the honeycomb, the rock-art formlings tend to have two parts; the center darkened by the cells containing developing bees, or brood, and the outside lighter because it is full of clear, golden honey.

These images were created by the bushmen, or San, who have lived in southern Africa for perhaps as long as 25,000 years. With their pointed chins, golden skin, wide cheekbones, slanted, almost Asiatic eyelids, and buttocks that seem designed to carry a stored wedge of fat, as a camel has a hump for times of food shortage, they are recognizably a distinctive people within Africa. When they were studied by anthropologists, and celebrated by the writer Laurens van der Post in the 1950s and 1960s, there were around five thousand still using their wits and prowess to feed themselves, at least in part, from their environment. Although these are not exactly the same race who made the ancient rock drawings of Zimbabwe and South Africa, who spoke a different tongue and became extinct in the nineteenth century, all the same, by studying both the living and records of the dead, particularly the bushmen and of other honey-hunting people such as the Mbuti pygmies of the Congo, it is possible to get closer to the artists of the past, and what bees and honey-hunting meant to them.

The Kalahari is not, at first sight, a well-stocked pantry. A vast expanse, covering one-third of southern Africa, it has no running water. Its sandy soil and interrupted pelt of bush grass, as dry and yellow as a lion's mane, is marked at points by the strange bulbous shapes of baobab trees, with barrel-like trunks, up to 30 feet in diameter, and branches that look like roots, as if the tree had been lifted up and stuck back upside down. This tree is a favorite site for honeybees to nest. It is a land that has to be read and tracked with the eyes of a lynx, the speed of a cheetah, and an astonishing amount of stamina.

Nineteenth-century European settlers were astonished that the bushmen could see objects they themselves could only perceive through a telescope. They spoke of a race of men in southern Africa who, it was said, were so quick and accurate in pursuit of a trail, they had eyes in their feet. There are still hunters in the Kalahari who can literally run down their prey, by following tracks for as long as eight hours.

Finding honey was a cross between hunting and gathering. First, you had to find the nest. This was done by the bushmen in the same way they would follow an antelope by its tracks: by the persistent pursuit of small signs. As well as recognizing places where the bees habitually nested, the bushmen would spot bees in the air, flying back to their nest, and follow the direction, picking up the trail from a succession of insects until they noticed some going in a contrary direction; at this point, they knew they had passed the location of the honeycomb and turned around. Some other Africans were deft enough to attach a short strand of cotton, a long white ox hair, or a piece of grass to the bees to make them more traceable. Tracked in the evening, the bees became more easily visible as dark dots against the setting sun, or with the last light catching their shimmering wings.

The ability to follow a series of fast-flying dots is extraordinary, but the human also had, and still has, an avian ally. The greater honeyguide (with the appropriate name *Indicator indicator*) is a bird with the exceptional ability to eat wax, which other creatures find indigestible. But it needs help in getting to the bees' nest, often inaccessible within a hollow tree, the cleft of a rock, or in the ground. Honeyguides use several sweet-loving mammals, particularly the honeybadger or ratel, to gain access to the nest, but the handiest mammal of all is man. When a bushman hunter first hears the whirring *cherr, cherr, cherr* of the honeyguide, he knows there is honey to be had. The bird flies a short distance. It calls again. The man follows. It calls. He follows. At each step, the bird's call becomes more insistent and the distances shorten as the two close in on the nest. The man also calls to the bird, perhaps using a snail shell, or his own form of bird call. When it finally arrives, at the end of this long sequence of call-and-seek, the bird hovers over the nest, before perching nearby, waiting for the man to finish the job. He hacks open the tree and the honey oozes out. It can take a whole day to find a nest, but just a few hours with a honeyguide. The man leaves a piece of comb containing the prized, nutritious bee brood. And the reward may not be so big as to sate the bird's appetite; otherwise, it might not lead the hunters on to the next pot of gold.

Once a tree was identified as a source of honey, the hunters might tie a piece of twine around the trunk, signaling it as taken territory. Such a honey tree might then have wooden pegs driven into the wood, to make it easier to climb up to the nest. Improvised ladders, made out of tree branches, were also propped against trees or rocks to reach honeycomb sites. Some tribes became adept at climbing trees either by wrapping their arms and legs around the trunk and shinning up or, if the tree has a larger dimension, by slinging a length of vine (or later a belt) around the trunk like an extension of the arms, to inch their way up to the nest. Another method was to climb a parallel tree and then climb over onto an otherwise inaccessible branch. Once up, the hunters hacked into the wood and pulled out the comb. Sometimes, smoke was used to pacify the bees, just as it still is today by hive beekeepers.

The hunters and their families feasted on the honey right away, gorging on the comb as a treat. The brood comb, containing the developing bee larvae, was especially desired; the anthropologist Colin M. Turnbull, in his book on the Congo, *Wayward Servants*, describes how the Mbuti honey hunters would warm the comb gently to make the larvae wriggle as they were eaten. In much of the Western world, there is a taboo against eating insects, but many cultures see insects as a useful source of protein, and have preferences for different ways of eating them. In Zambia, for example, the unsealed brood comb is a delicacy while the sealed comb is left uneaten, perhaps because of the risk of eating nearly formed bees.

While the hunters would devour some of the comb immediately—gleefully licking their arms and fingers as their upper bodies became sticky with honey—they kept some to be shared. Dividing up food among a group lessens the risk and fear of starvation and has a profound significance for bonding. Typically, among African tribal peoples, the successful hunter is not boastful of his kill. He might at first deny to the group, on his return, that he has had any amount of success, and the fellow hunters might tease him about the smallness of his catch as they heave it back to the others left in the camp. Successful hunting and sharing food are part of survival and they are also the cornerstone of social cohesion within a group. Not sharing food verges on the taboo.

Hunting-gathering, far from a random and desperate lurch between life and death, is a skillful way of life that requires coordination and knowledge in order to succeed. Land is not so much property as territory, a given area that can supply the needs of a certain number of people. The bushmen, like other hunter-gatherer tribes, divided the year into two halves: a dry season, when they gathered together in large groups of perhaps a hundred, and the wetter season, when they broke into smaller groups of about thirty and moved around to find food in a highly organized but nomadic pattern. Such times spent in a smaller unit were an opportunity for closer bonding and any disagreements could be resolved in the larger gathering, before formations of new, smaller groups for the next hunting season. This pattern of fusion and fission is still a typical part of the hunter-gatherer way of life. Mbuti pygmies in the equatorial rain forest will go on twelve-day honey-hunting expeditions, collecting more than 66 pounds of honey per day, and a similar process of reforming and bonding takes place on these trips.

In traditional African hunter-gatherer societies, men hunt and women gather. When anthropologist Richard Lee studied one group of bushmen, the !Kung in northwest Kalahari in the 1960s, he calculated the economics of survival. The women of the tribe could walk for fifteen hundred miles a year, slowly collecting nutrients for more than three-quarters of the diet, spotting small wisps sticking out from the ground that they would follow down, digging with their grubbing sticks, to reach the nutritious tubers below; patiently picking berries, or sucking water through a hollow stem to be collected in empty ostrich eggs, which are plugged with grass and left buried in the ground for times of need. Such gleanings—small in themselves—built up to a substantial percentage of the diet and could be given just to their immediate family, within the group.

But the bounty of the male hunters, while it adds up to far fewer nutrients, carried far more significance within the group. As honey was mostly hunted, as opposed to gathered, it had meanings other than nutrition attached to it. Laurens van der Post, whose search for the bushmen is described in his book *The Lost World of the Kalahari*, wrote of the bushman: "Typically, he raised the search for honey into a kind of sacramental adventure." Bushman songs show how much the bounty was prized and desired. One hunter's song asks the Great Father to let him find sweet roots and honey, and water to drink; another is about people carrying honey and flesh back home to their hungry women who need food.

The two-month honey-hunting season of the Mbuti in the Ituri forest of the Congo is described by Colin Turnbull as a time of festivity and magic. Special songs and dances relate to bees and honey. In one, the men sing a honey song while the women buzz. The men "find" the honey and the women "sting" them with burning embers. In another ritual, a woman blows through a special stick into the ears of the men, making a sound that they say resembles the sounds of the brood eating honey in the comb, and this is meant to help guide them to the bees in the trees above.

People misinterpret ancient rock carving by seeing it simply as daubed representations of surroundings, the equivalent of Sunday watercolors. By joining these enduring images with observations of tribal customs, art historians and anthropologists are beginning to understand what lies behind them, and therefore something of how early humans related to their world.

~

THE DIVISIONS WE MAKE, between animal and human, and between visible and invisible, are more rigid than those made by the bushmen. For them, the spiritual and material dissolved into each other; the God-like figures in bushman stories assume animal forms, rather as Zeus does in the stories of the ancient Greeks. Some rock images show animals disappearing into the clefts of rocks, and David Lewis-Williams, an expert in such artworks, has interpreted this by seeing the rock as the interface between the physical and invisible worlds, with the animal able to pass between them.

Humans could also move between the two worlds by becoming more like other animals, a feat performed through trance. Such trance states were achieved not by drugs but by ritual. For example, bushmen danced around a fire, close together, their clapping, songs, and movements beating louder and faster until a state of delirium was achieved. Individuals might then fall to the ground. In this state, the human could reach the spirit and power contained in animals, and become closer to them. Rock carvings reflect this transition between man and animal; a classic example is the figure with an antelope head and a human body.

This leads back to the mysterious geometric images of nested curves. It seems an extraordinary coincidence that the same image should occur in both southern Africa and the Iberian peninsula. Could some sort of tradition have crossed continents to pass from one people to another? Or is there a different explanation?

The state of trance can also be reached through psychotropic substances. Scientists have recorded that humans taking hallucinogenic drugs such as LSD under laboratory conditions start to "see" a succession of glowing geometric patterns that pass across the eyes. These are the same for tribal peoples performing rituals as they are for research subjects taking drugs in laboratories, suggesting that they have a common neurological origin. After seeing these geometric patterns, called entoptics, the human mind, with its constant need for explanation, starts to interpret them, fitting them to images with which they are familiar.

Lewis-Williams thinks a bushman in a state of trance might see this pattern of nested semicircles and then associate it with a bees' nest. The rock artists may have been shamans who interpreted what they felt and saw in a state of trance, and then made the images we see on the granite rocks of southern Africa. Honey was, for them, the miraculous product of a sacred insect; the marks they made were part of its magic.

Hunter-gathering was a way of life our ancestors pursued for 99 percent of human existence. The tiny proportion of mankind who today live in this way still attach enormous significance to the finding and eating of food. Honey continued to be hunted, whether by medieval peasants in the forests of central Europe or by today's Gurung tribe of Nepal in the foothills of the Himalayas. But other ancestors were about to make a seismic shift, from hunting-gathering to agriculture, and with it came beekeeping.

CHAPTER THREE

ORGANIZATION AND MAGIC

While humans were hunter-gatherers, the honeybee inhabited the dark cavities of rock crevices and hollow trees; but as we began to settle, to grow crops and rear animals, new spaces were created that bees could inhabit. This probably arose by accident rather than design. The water pots and baskets made by Neolithic humans to aid their fledgling agriculture would have been perfect spaces for bees; it isn't hard to imagine roving swarms settling in ones left lying around. It is likely that humans would then raid these improvised nests, since it was easier than hunting down honeycomb in the wild. The next step would have been to create spaces specifically for bees: why not encourage them to settle where the honey could be collected more easily? This seems to be the best explanation as to how beekeeping began. The insects were still wild, just encouraged to live closer to humans.

Through the papyrus records, paintings, and sculptures that were preserved in the hot, dry climate of Egypt we know something of how the first farmers kept bees. Beekeeping was one of the many lasting achievements of the ancient Egyptians, and the pattern they set remains in many ways essentially the same today as it was five millennia ago.

Egypt is a land of two contrasting parts: the rich, fertile black soil of the alluvial plain, which you still see clinging to new potatoes in greengrocers, and the arid red of the surrounding desert. The green strip of vegetation that grows on the black soil is threaded on the Nile through the center of the desert and is composed of the fine, rich mud deposited by the river. Prior to the building of the High Aswan Dam, this land was brought back to life and replenished each year by floodwaters, as well as sustained throughout the year by irrigation. In a land where the rainfall can be sparse, or even nonexistent, and where the sun beats down from a cloudless sky, this river has always been the region's lifeblood. No wonder that ancient Egyptians called their land the gift of the Nile.

The fertile green strip along the Nile provided plenty of foraging for the bees. Pollen grains found in ancient honeys include those crops grown for animal fodder, such as alfalfa, clover, and chickling pea. The fruits and vegetables grown for human consumption, such as lemons and beans, provided food, too, for the honeybee, in the form of nectar and pollen. The ancient Egyptians also loved their gardens; one of their prayers was that, after death, you would sit in the shade and enjoy the fruit of trees you had planted. The garlands discovered on the mummified body of Tutankhamen (King Tut) indicate the range of plants prized by the Egyptians that also provided nectar for honey, such as cornflowers and wild celery.

Settled civilization began on riverbanks and its roots were in agriculture. The Egyptians were among the first to centralize and organize food production, including beekeeping. As well as irrigating and cultivating the soil, they bred cattle, farmed pigs, had primitive incubators for eggs to create the first poultry farming, and stored grain. There was enough food produced in this successful agricultural system to feed the workforce that built the pyramids. Life was less hand-to-mouth—more planned than before. Just as scribes counted the peasants' produce, stored it in state granaries, and distributed it back to the people, there was a specialized honey scribe to count jarfuls of honey. These early civil servants monitored every part of this "black soil economy," leaving a fascinating record of daily life. Their papyri tell us how Ramses III provided 20,800 jars of honey as an offering to the Nile god. On a more domestic scale, it is through them that we know about the marriage vow that includes the line "I take thee to wife . . . and promise to deliver to thee yearly twelve jars of honey."

To have such large quantities of honey, the ancient Egyptians must have gathered in regular supplies. Honey-hunting continued in the desert fringes of the fertile area, where specialist bee hunters, called *bityw,* undertook expeditions, alongside collectors of turpentine resin and archers to protect them from the dangers of this hostile environment; but the bulk of their honey must have come from the more reliable source of hives.

The earliest archaeological evidence of hive honey is from a stone bas-relief in the sun temple of King Ne-user-re at Abu Ghorab, dating to around 2400 BC. It shows a sequence of production, from hive to sealed pot. Although one end of the sculpture is partly destroyed, it is still possible to see the edge of the cylindrical hives. A hieroglyph above the image indicates that a current of air is being used, probably meaning the smoke used to pacify the bees. The man holding the smoker is obscured by damage but other images from later periods show this use of smoke quite clearly, with beekeepers holding pots emitting stylized flames toward the open ends of hives. The image in the sun temple next shows three men pouring honey into large containers, and then another setting a seal on some closed, round pots. The hieroglyphs continue: "filling, pressing, sealing of honey."

A very similar sequence can be seen on a wall painting in one of the tombs of Prime Minister Rekhmire, at the West Bank of Luxor (1450 BC). On the carved wall, a bank of three hives is clearly visible this time, with a beekeeper holding a burner toward the hives while another takes a comb out of the middle hive to add to the combs already piled up in two dishes. Other men in the scene have jars, large containers, and sealed dishes—the vessels the honey must have been kept in. The hives are a purplish gray, the color of dried, unfired Nile mud.

Dr. Eva Crane is an international bee and honey expert. Over the course of fifty years she has traveled to sixty countries, riding by sled in Alaska, boating in the Mekong delta, descending a rope to inspect hives on a rock ledge in the Pyrenees—going anywhere and everywhere in order to meet beekeepers around the world. Her quest began when she was given a box of bees as a wedding present in 1942, at a time when honey was a boost to the dull wartime diet. A university lecturer with a PhD in nuclear physics, she began to make more inquiries about bees, and realized there was no coordinated way to learn about beekeeping around the world. She set up such an information pool, and was director of the International Bee Research Association, now based in Cardiff, between 1949 and 1983. Among her many travels, described in her memoir *Making a Bee-line*, and her magnum opus, *The World History of Beekeeping and Honey Hunting*, Dr. Crane visited modern Egypt to see if current beekeeping could throw light on the methods of the ancients. Her observations help join the fragments that we have inherited through archaeology. She saw that modern beekeepers were using cylindrical hives made of dried mud, which they piled up in great banks, each layer helping to support the one above. In 1978, while traveling through the Nile valley just north of Assyut, Dr. Crane saw an estimated ten thousand such hives in just 17 miles. The cylindrical form of the hives and the way they are placed on top of each other is strikingly similar to the evidence of ancient art.

The method still used to make these traditional hives could well date back to ancient times. A layer of mud is smeared on a mat made of straight plant stalks such as reeds, rolled up and left to dry; then the mat is cut away to leave a long cylinder. Both ends are then

Painting in Rekhmire's tomb in Luxor, showing hives being smoked and honeycombs removed and packed into jars.

covered with mud discs, with a hole in one end for the bees to come and go. The beekeeper opens the other end of the hive to collect the honey-filled combs. Dr. Crane also saw cow dung, a common form of fuel in the Middle East, being used to smoke bees; this could easily have been the method used in ancient Egyptian beekeeping.

Other current observations give us clues as to the methods of the ancients. An eighteenth-century traveler in Egypt described how hives were moved gradually down the Nile on boats, leaving Upper Egypt at the end of October. The hives were put on rafts and the bees went to successive fields in flower as they were moved northward toward the Mediterranean. By the start of February, this traveling workforce reached Cairo, where the honey collected on the journey was sold. Migratory beekeeping is a major source of income for today's beekeepers in many parts of the world, who take their hives around, following the crops currently in bloom, their efforts paid for by the farmers who want their crops pollinated by the bees.

The process of pollination was unknown to the ancient Egyptians, but it is not hard to imagine this boat-based society moving their hives about in order to get a good crop of honey. There is some archaeological evidence to show that beekeepers and farmers worked together from the start—though the relationship was not always an easy one. A pitiful, desperate tone pervades a third-century BC petition from beekeepers calling for donkeys, urgently, to bring their hives back from the fields before the impatient peasants, wanting to press on, flooded the land and destroyed the hives: "Unless the donkeys are sent at once, the result will be that the hives will be ruined."

THE MECHANICS OF BEEKEEPING in ancient Egypt are not all that we know; we also know something of how honey and other bee products were used in the household, both for practical and for other, more mystical, purposes.

The diet of ancient Egypt has some striking similarities with that of today. The grain, honey, and cattle shown in the ancient pictures became the bread, beer, beef, and honey-rich cakes on their tables, as on ours. Food differed according to the rank and need of different parts of society. The messenger and standard-bearer of King Sethos I ate bread, beef, wine, sweet oil, olive oil, fat, honey, figs, fish, and vegetables, a combination that could easily suit the modern tastes of a contemporary civil servant eating a Mediterranean diet. Quarry workers at that time got a more functional, though nutritious, ration of 4½ pounds of bread, two bundles of vegetables, and a piece of roast meat.

Honey was a food of the rich—the poor sweetened their food with dates—and its high status meant it was celebrated in poetry at feasts. The daughter of Prime Minister Rekhmire had a song sung at a party: "The little sycamore, which she hath planted with her hand, it moveth its mouth to speak. The whispering of its leaves is as sweet as refined honey."

There are honey cakes depicted in Rekhmire's tomb that can still be found in different versions today. The scribes record large quantities of honey being used for such cakes as offerings, which shows how much they were valued. The Agricultural Museum in Dokki has an ancient honey feast cake shaped into the form of a human, with a head and arms, possibly, Dr. Crane suggests, an early form of a gingerbread man.

Honey and beeswax occur in many other parts of ancient Egyptian daily life. Wax was used to hold the waves of the elaborate wigs used by both men and women. The long braided and sculpted tresses, some composed of more than 120,000 human hairs, are reminiscent of eighteenth-century wigs, or even the hair extensions of today and the sculpted forms of dreadlocks. Bee products have been found in the cosmetic boxes filled with kohl and other colorants, lotions, and beauty potions. Honey was, for example, included in a compound of powdered alabaster, powdered natron (sodium carbonate, much used in the mummification process), and salt, which was supposed to smooth wrinkles.

Honey was also used for more personal treatments. A mixture of honey, herbs, oils, and onions was applied to a woman's vagina to try to stop a miscarriage, while crocodile feces, honey, and saltpeter were used to stop conception in the first place (perhaps such a substance would work by putting you off sex altogether). Dr. Crane was told in 1993 that cotton soaked in lemon and honey was still used as a contraceptive in Egypt.

The fact that honey was added to so many of the medicines of ancient Egypt may have been a case of sweetening the pill when remedies included the likes of fingernail dirt and mouse droppings.

Ingredients like these were supposed to be repulsive to the evils inhabiting the body. Putting aside such deliberately disgusting treatments, and such hocus-pocus as curses for catarrh, we still have lessons to learn from the physician-priests of ancient Egypt, who were much admired by the ancient Greeks, usually seen as the founders of medicine. Honey was certainly put in dressings for wounds—an effective measure, as modern medicine was later to prove. When honey is mixed with body fluids, it produces hydrogen peroxide, which inhibits the growth of bacteria. For similar reasons, honey was—and still is—used for gut problems, and even for tooth decay. The remains of ancient Egyptians reveal that they had strikingly bad teeth—perhaps from having to grind down the coarse grains—and a mixture of honey and herbs would be packed into the mouth to fight infection. The modern mind may scoff at such an idea, but it was, in fact, a perfectly good remedy, and is starting to be used again today.

ANCIENT EGYPT casts a spell on us with its strange mixture of practicality and mysticism. Its priest-physicians were using a mixture of medicine and magic. Many papyrus records contain formulae that sound like plausible prescriptions, with a tone of precision and correctness, before suddenly veering off into spells—rather as if a solidly reassuring doctor's note had vaporized into sparkling ectoplasm. This mixture of knowledge and the metaphysical intrigues us today; in the same way, we admire the engineering feat of the pyramids and at the same time feel alienated—and fascinated—by their purpose as launch pads into eternity.

The mummy is the obvious physical embodiment of this cult of the afterlife. For a high-class embalming, first the entrails of the corpse were removed, preserved, and wrapped in separate packets (the brain was discarded as irrelevant). The rest of the body was then dried by putting it in natron for forty days. It was washed, purified, and dried, the skin rubbed with honey, milk, and ointments. Because honey has preservative qualities, it also played additional parts in some burials. The corpse of Alexander the Great was said to have been covered in honey, on his own instructions, to stop its putrefying until he was buried. In *The Mummy*, the British Museum's Egyptologist Sir Wallis Budge described the discovery of a honeyed corpse. A jar of ancient honey was discovered in a grave by some treasure seekers. One was dipping his bread into the honey for a taste when he noticed a hair: "[A]s they drew it out the body of a small child appeared with all its limbs complete and in a good state of preservation; it was well dressed and had upon it numerous ornaments."

Such a ghoulish incident, like the curses on the doors of tombs in musty pyramids, is part of this frisson of ancient Egypt. Beeswax figures are similarly chilling in this effect. They were said to prefigure what you hoped would happen. One papyrus from around 2830 BC, for example, tells of the spells of a man called Aba-aner. From a box of magic, he produced a wax crocodile and put it into the water where his wife's lover was bathing. It came alive in the water and dragged the man down into the deep. This grisly tale shows a belief in the power of beeswax figures to move in and out of reality.

Medicine and magic both drew upon the idea of *haka*, the term the ancient Egyptians used for the power that came from creation. A potion, a prescription, a prayer: all turned toward the same source. This ultimate power could be embodied in animals and statues, and they figure in the amulets, or magic charms, that were carried about for good fortune. They could take the form of insects, sometimes breathing magic into a close observation of nature. The scarab beetle rolls a ball of dung into a hole and lays its eggs inside to give them warmth and food; the Egyptians depict the scarab as

rolling the life-giving sun across the sky each day. The ancient
Egyptians believed bees originated as tears of the Sun god, Ra,
merging the two crucial substances of water and sunlight in the
body of the bee. The Salt Magical Papyrus records: "When Ra
weeps again the water which flows from his eyes upon the ground
turns into working bees. They work in flowers and trees of every
kind and wax and honey come into being." As we've already seen,
honey was sacred enough to be a suitable offering to the Nile gods;
sacred bees were also part of the Egyptian's cult of the afterlife, with
honey one of the funerary foods left in the tombs of the mummified dead, as sustenance for the next world. The Opening of the
Mouth ceremony was performed on mummies so they could "eat"

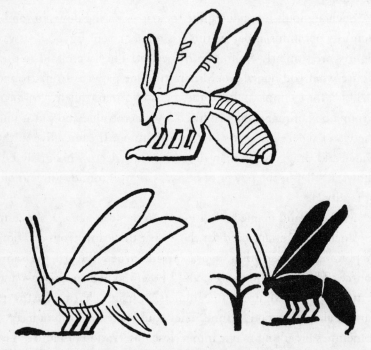

Old Kingdom (top), Twelfth Dynasty, and Sixth Dynasty bees.

these funerary foods. In the words of one such ritual, the Kher-heb, recorded by Wallis Budge, the priest foresees the new life of the soul: "Going about as a bee, thou seest all the goings about of thy father."

Bees can be seen all over the monumental remains of ancient Egypt for another reason. The symbol of Upper Egypt was a sedge plant, and the symbol of Lower Egypt, a bee. The two are unified in the hieroglyphs as part of the title of the King, or Pharaoh, of Egypt. Ancient Egypt was a unified, organized civilization in a world of warring tribes, and the titulary of the bee and the sedge plant is one of significant unity. It may be that the coming together of Upper and Lower Egypt is partly mythical, a concept rather than a matter of history. Either way, the hieroglyphic bee in the Pharaoh's title represents powerful concepts. The bee was part of the Pharaoh's godlike glory and its honey was offered to him and to other gods; the honeybee had become part of the divinity of the land.

CHAPTER FOUR

FOOD OF THE GODS

Kronos, the father of the Olympian gods, clung to power by infanticide: warned that he would be dethroned by one of his children, he took to swallowing them at birth. After the birth of a sixth child, his wife, Rhea, instead gave him a stone wrapped in swaddling clothes, and took the latest newborn into hiding in a cave on Mount Dicte in Crete. Out of sight, the baby Zeus, destined to be the god of gods, was still in danger of detection from the insistent sound of his crying. Young guards hid the noise by clashing their spears and armor. As well as being one of the few sounds that could possibly override a baby's wails, clashing brass was believed by the ancients to attract swarms of bees; insects came to the cave of Zeus and settled, and the infant deity was nourished on milk and honey. So it was that honey became a food of the gods—and poetry added yet more luster to its gold.

In a story, anything can happen. The classical writers elevated deities to powerful, Olympian heights; on the other hand, they could dash mortals down with obliterating fate. What gave them the power? Honeyed words.

Bees hovering near the lips of a newborn was an omen that the child would grow up with a mellifluous tongue. Bees are said to have flown near the milky infant mouth of Virgil, the greatest poet of Classical Rome. The breath of life is certainly present in his many references to them; some scholars and apiarist readers believe he himself was a beekeeper, such is the detail and freshness of his observations.

The fourth and last book of Virgil's great poem on the art of farming, the *Georgics*, is much about honeybees, describing their collective work, frugal ways, organization, and obedience to leadership. A lesser work on this theme might have had a simplistic ring of stern, Roman virtue, but instead the poetry hums with life in all its complexity.

Virgil spent his adult life in the countryside near Naples, away from the politicking and power-grabbing of Rome during a particularly tumultuous period. Nearly a third of his life had the backdrop of civil war. The *Georgics* were begun about six years after the assassination of Julius Caesar in 44 BC, and the moral, peaceful rural life of this poem communicates, subliminally, a reaction to the

Ancient Greek amphora depicting sacred bees stinging intruders in the cave on Mount Dicte.

trauma of war beyond its pages. The Latin name for the poem, *Georgica*, translates as "what concerns the man that works the earth," and was about such smallholders, rather than the large Roman estates that used slave labor.

Nature runs through the lines of *Georgics, Book Four,* like a stream. You see and smell the flowers that attract the bees—the wild thyme, the "rich breathing" savory, the bank green with celery, the lime blossom, willow, saffron, and lily. You recognize how the bees "hurry from the hive, all helter-skelter" (this translation is by Cecil Day-Lewis, who was himself working against the backdrop of the Second World War); the "vague and wind-warped column of cloud to your wondering eyes" of a swarm; the bucolic image of an old smallholder, happy as a king when his small, poor land can yield roses in spring, apples in autumn, and frothy honey squeezed from the combs.

The poem partly poses as instruction, and it is fascinating to read what Virgil thought of bees and how they should be kept. He encourages the beekeeper to keep his hives near water, so the bees can drink, and says the hives must be sheltered from rough winds. He emphasizes the importance of the hive's ruler, and perhaps this commanding figure is meant to suggest, on a political dimension, a leader who could bring unity to the warring factions of Rome, spinning out of control after the murder of Julius Caesar.

The poem tells the beekeeper to site the hive away from lizards, the "sinister tribe of moths," and from birds that eat bees. Some of this is all too recognizable today. As well as the moths that can devastate a colony, birds with strong beaks, such as the woodpecker, have been known to break into bees' nests in the wild, and there are twenty-five species of bee eaters, including one occasionally found in Britain, that can catch bees on the wing in their beaks and pull them over a hard surface to extract the venom before eating them. (Scientists dissecting spine-tailed swifts in the

Philippines found one bird with the remains of nearly four hundred bees, the bird's mouth and gizzards barbed with scores of detached stings.)

Virgil mentions the three kinds of bees in the hive. His probable source was Aristotle (384–321 BC), who wrote extensively about bees in his natural history writing, particularly in *Historia Animalium*. As well as noting there are different sorts of honeybees, all with different roles, Aristotle's book describes how the bees collect the juices of the flowers in their stomachs and take them back to the hive to regurgitate into the wax, and that this liquid gets thicker with time. Aristotle saw that there were hairy worker bees inside the hive and smoother ones outside (the worker bees become less fluffy with age), and that the new ruler could kill off others that emerged from other cells. He even noticed how the bees danced across the face of the comb, though he did not understand the meaning of this (we now know it is a means of communication).

Some mysteries remained. Many mused on the reproductive methods of bees. Aristotle was especially interested in where bees came from, though he never reached a satisfactory explanation, pondering whether the young were collected from flowers, olives, and reeds. The strangest belief of all, maintained for many centuries, was that honeybees generated spontaneously from the carcass of an ox. Credence in this idea of "ox-born bees" continued right up to the time when a certain Mr. Carew reported this feat of reproduction in Coventry in 1842. Virgil praises the worker bees' abstention from sexual intercourse, an escape from the mess of passion and the pain of birth. "How to get bees from an ox" appears in his *Georgics, Book Four,* almost like a recipe: in the springtime, you must take a two-year-old ox into a small house with four windows, stop up its nostrils and mouth, bludgeon it to death, and leave it in the room, along with cassia, thyme, and branches.

Virgil's description of how bees then poured out of the rotting flesh, like a throbbing attack of arrows, is so like the pulsating release of flies from the maggots feeding on carrion that this could be one natural explanation for this curious belief. The honeybee does not settle on meat; but it easily could have been confused with the drone fly that lays its eggs in decomposing carcasses. Another possibility is that bees do nest in skulls, and among other bones, in places where shelter is rare, such as the Egyptian desert. This could also be the origin of the Old Testament story of Samson (Judges 14) when he asks the riddle "out of the strong came forth sweetness"; the image of a swarm of bees and a lion remains on tins of Tate & Lyle's golden syrup.

In Virgil's *Georgics,* there is a more important interpretation than the literal truth of the ox-born bee: this is the idea that life regenerates. A swarm of bees is a beautiful, visible embodiment of reproduction; moving through the air in its dark cluster, it is a living symbol of how life moves on. In the beliefs of the ancient classical cultures, this concept was reinforced because the best swarms happen in spring—a time of the annual rebirth of the world and the continuation of life after the death of winter. Whether a swarm of bees comes from an ox or a hive, it is the start of new life, flying toward the future. Hardly surprising, then, that bees came to be portrayed as special creatures that could move between life and death, between the world and the underworld, between humans and the divine. In ancient Greece, bees flying through the cracks of rocks were thought to be souls emerging from the underworld, just as the ancient Egyptians believed the insects to be human spirits that could fly anywhere.

One of the most moving passages in *Georgics, Book Four,* describes the bees' blending of the natural and supernatural. It connects these flying particles of heaven to the way in which all of nature—including man—exists (the translation is Dryden's):

Ox-born bees in an engraving illustrating John Dryden's translation of Virgil's
Georgics, Book Four.

...some have taught
That Bees have Portions of Etherial Thought:
Endu'd with Particles of Heavenly Fires:
For God the whole created Mass inspires;
Thro' Heav'n, and Earth, and Oceans depth he throws
His Influence round, and kindles as he goes.
Hence Flocks, and Herds, and men, and Beasts, and Fowls
With Breath are quickn'd; and attract their Souls.
Hence take the Forms his Prescience did ordain,
And into him at length resolve again.
No room is left for Death, they mount the Sky,
And to their own congenial Planets fly.

Life comes, briefly, and then is reabsorbed back into the heavens and never dies. I asked Willie Robson, on the Northumberland heather moors, about the life span of a colony. The bees? They're immortal, he said.

WHO WERE THE Greek and Roman gods? In his exploration of Sicily, *The Golden Honeycomb*, Vincent Cronin described the ancient world as a twilight zone where legend and history met. The stories of mythology gave gods and goddesses human traits, raised up on Mount Olympus; but they were, for all their glory, ultimately akin to men and women, with characteristics and foibles we recognize. This was the time when humans began to civilize the Western world, to form it to our own will. Mythology was a way of projecting us onto a larger scale, with all the possibilities within the reach of our imagination—and all our mistakes, too. Greek and Roman gods certainly feel more familiar to us than the mysterious deities of ancient Egypt, worshipped from afar with fearful awe.

The Golden Honeycomb is based around Cronin's quest in the

1950s to uncover the origin of one such myth, the tale of Daedalus, the legendary master craftsman and inventor who supposedly flew to Sicily from Crete, believed by the ancients to be the origin of bees and beekeeping. On the way, Daedalus's son Icarus flew too near to the sun, and the beeswax holding his wings together melted, plunging him into the sea, where he drowned. But Daedalus made it to the island and here was said to have wrought an exquisitely realistic golden honeycomb. Did it actually exist, or was it a metaphor of some sort? Cronin travels around looking for clues. He goes to Greek ruins, such as Silenus, named after wild celery, which is a good source of nectar for bees; he reads the sweet words of poets; he writes of how nature was connected to divine forces. In Taormina he sees the spring, with its lavish surge of life and the flowers that provide so much nectar for the bees—and the reassurance that humans, too, will be able to eat again after the winter. Was this vital expression of Mediterranean nature in some way connected to the origin of the golden honeycomb?

No conclusive answer could be found; rather, Cronin's search gives him a reason to explore successive cultures, and the ways in which honey played a part in them.

Alongside oil or wine, honey was one of the libations for the dead in the classical era. Achilles put jars of oil and honey by the funeral pyre of his friend Patroclus, so the food of this life could be enjoyed in the afterlife. Honey was offered to the gods because it was a product of both earth and sky; it was believed the bees gathered the juice that had fallen from the heavens and collected in flowers, and honey was, therefore, an appropriate food with which to commune with the otherworld. Such beliefs in the sanctity of bees continued. When Cronin was writing in the 1950s, Sicilian newlyweds arriving home from the church for the bridal feast were given a loving spoonful of honey to share. Death, life, mythology, and love: honey slid into them all.

~

THE ANCIENT GREEKS and Romans were the first serious observers of the natural history of the honeybee. After Aristotle, the next most significant author on bees was Varro (116–27 BC), the greatest Roman scholar of his day. His *Res rusticae* was written at roughly the same time as Virgil's *Georgics, Book Four*. This practical, useful text goes into subjects such as the different types of hive, evaluating those made from wood, bark, earthenware, and reeds. He writes about the plants that are good for honey production, such as thyme, beans, and the succession of blooms between the spring and autumn equinoxes. He even mentions the economics of beekeeping, telling of two Spanish brothers who ran a successful apiary on just half an acre of land. He writes, too, of the value of propolis; in Rome's Via Sacra, it was more expensive than wax. Propolis is the sticky, dark "bee-glue" gathered from buds and the bark of trees, with which the bees seal up their hives (the word comes from the Greek "before the city," meaning it surrounded the city, or colony, of bees). Even at this early date, it was valued for its bacteriocidal and fungicidal properties, and was used by doctors in the classical world to treat ulcers and tumors.

Another Roman writer, Pliny the Elder (AD 23–79), noted that propolis could also draw out stings and foreign bodies. As for understanding the bees, he mentions a consul who had hives made of translucent horn so that he could watch the new bees emerging from their cells. It sounds like a primitive version of an observation hive, though not a particularly clear one. Pliny's own questions and observations could also be opaque, if poetic. Was honey the saliva of the stars or the sweat of the sky, he asked? A little more on target, he noted how honey thickened and was covered with a skin that was "the foam of the boiling." This kitchen metaphor, of a liquid reducing through evaporation, is analogous to what happens to honey as it thickens through the fanning of the insects' wings (though the "skin" is made of wax).

There was still a misunderstanding about the origin of beeswax, which was thought to be a secretion collected from plants rather than a product of the bee itself. Whatever its source, the wax itself was much used on a practical, everyday level. Wooden boards, coated with wax, were reusable writing tablets. Wax was also used by craftsmen, real and mythological, for joining objects together— unsuccessfully, in the case of Icarus's wings and more successfully, with Pan's pipes.

In the artistic field, statues were cast using the *cire perdue* or "lost wax" method. Malleable beeswax was modeled and then covered in clay; this was then heated so the wax melted, leaving a cast in one piece into which molten metal could be poured. With other methods, the cast had to be cut into pieces; with the lost-wax method, a complete mold could be made. Wax busts were also sculpted of famous people, a technique that is still used. Besides this, the flaws in dodgy classical statues could be hidden with beeswax. If a statue was *sans cere*, or without wax, the seller was an honest dealer or, as you could say, sincere.

Writing about bees and honey continued, but more in encyclopedic collections of received wisdom than in fresh observation. Fifty years after both Virgil and Varro came Columella, an army officer from Cadiz who retired outside Rome. His book on agriculture, *De re rustica*, written around AD 60, methodically runs through many aspects of the hive, from the races of bees to the extraction of beeswax. Palladius, writing in the fourth century AD, gives a month-by-month account of the honeybee; his book was translated into English in the fourteenth century. Classical authority was set to remain largely unquestioned for a thousand years— then the chief work of the Greek physician Dioscorides, *De materia medica*, written in the first century AD, was a crucial text on botany and healing through the Middle Ages and was used right up to the seventeenth century.

~

THE ANCIENT GREEKS and Romans already distinguished between different kinds of honey. Dioscorides said Attic honey from Greece was the best, with that from Mount Hymettus the very best of all. Honey from the islands of the Cyclades was next best, followed by that from Hybla in Sicily. They prized the distinctive quality and flavor of one particular honey made from the thyme that covered their hills and mountainsides. The herb imparts a special fragrance to the honey, giving it a unique taste of place. I associate thyme with the sinuous brown honey that you drizzle on sharp, white sheep's-milk yogurt for breakfast in Greece. After a trip to Sicily to see whether the honey culture of the ancients continued in this area today, I can now picture it among the limestone gorges of what is now called Monte Iblei, formerly Mount Hybla.

Sortino is a hill town on Monte Iblei where forty beekeepers still make at least part of a living from the nectar-rich slopes of their surroundings. At certain moments of my visit, the past millennia seemed to vanish. When I met Paolo Pagliaro, a sixth-generation beekeeper in his sixties, he almost immediately quoted Virgil to me, in Italian, his light blue eyes and youthful face lighting up as he spoke: *"Non vi è miele piùdolce di te, o miele ibleo!"* (There is no honey sweeter than you, o Iblean honey!)

Every October Paolo helps run a great honey festival, *Sagra del Miele,* which now attracts more than 65,000 people, and there runs a competition between honeys from all over the Mediterranean. I tasted his dizzying assortment of pots, from a surprisingly floral thistle honey to one that was darkly savory, almost like licorice, which turned out to be rose honey and cost $170 per pound. For all these novelties, I was most interested in trying his local varieties: the delicious, runny thyme, which was dark as a polished nut; the wild-flower honey with its up-front sweetness jolly as a child's painting of bright blooms; and the subtly floral orange blossom, which was slightly waxy in texture.

Evidence of the tradition of beekeeping on Mount Iblei came in different forms. One church had a collection of beeswax models embodying ailments such as broken legs, which reminded me of the effigies of ancient Egypt, though here in a Christian form. Paolo showed me his family's ferula hives. Right up until a generation ago, the beekeepers used such hives, which were made with the light, strong stems of a giant fennel threaded onto a wooden frame, using not a single nail, nor a single element that couldn't be gathered from the surrounding area. Such hives were mentioned by Varro and Columella; some ancient beekeepers thought them better than those made of pottery. Plant materials were said to be lighter, less break-able, and better at keeping the bees cool in hot weather. The small, rectangular hives would have been stacked in piles of eight high by twenty wide. You could get around 6 pounds of honey from one, a very small quantity compared to a modern hive. The ferula hives had that simple, handmade quality that is just a short step from the soil. When Paolo was young, he would move them around the coun-tryside on foot, pulling the cart himself, or with a horse.

Followers of Pythagoras, the Greek philosopher of the sixth century BC, believed you should breakfast on bread and honey every day for a disease-free, long life; while the philosopher Democritus (460–370 BC) advised that if you wanted longevity, you should moisten your insides with honey and your outside with oil. I asked Paolo about the longevity of honey eaters. Yes, he said, his grand-parents had both lived into their nineties. Paolo himself had a vigor and lightness undimmed by age.

~

NEAR SORTINO is Pantalica, a winding, flower-strewn limestone gorge with a Bronze Age necropolis of around five thousand tombs. These small, square-fronted caves were probably gouged out of the limestone with tools made from hard volcanic rock. The regular, simple entrances almost resemble large television screens or modernist architecture. The necropolis was certainly in use around 1200 BC, around the time archaeologists believe Troy was besieged. One later traveler, ignorant of Pantalica's origin, speculated whether these holes in the rock might have been carved for giant bees. When the gorge was properly explored by archaeologists in the nineteenth century, inside the tombs they discovered that bodies in the caves had been buried in the fetal position; in some, the heads rested on low stone ledges as if on pillows.

As I wandered down the gorge on a rubbly mule track, the scents of flowers arrived like snatches of birdsong. At the bottom was a pool of clear water and a New Ager wearing a rainbow T-shirt sitting on a rock and filling his pipe. He had come to a place of the utmost peace. In the stillness, the loudest sound was the river, like a rush of wind below my feet. It was not always so peaceful. Next to the young man's rock was a ruined mill where explosives had once been made using bat excrement from a nearby cave. Bees now massed on the ivy that had crawled all over the stones of the ruins. When Vincent Cronin descended the gorge on horseback from Sortino in the 1950s, he found wild honeycomb in the cleft of a rock, like a seam of gold; the climax of *The Golden Honeycomb* comes as he sees its swirls of honey-filled cells, the sun reflecting a hundred thousand suns in prismatic, reflected light. This was the end of his quest: not the myths, or the ruins, but the timeless glory of the honeycomb itself.

I found no wild honey in Pantalica, but the gorge was still full of hives. The honey plants of Mount Iblei succeed each other as the year moves on: the early spring almond blossom; the orange and lemon blossoms that make one of the major Sicilian honeys; the wildflowers that go into millefiori; the nectar-laden native oaks, which once covered the countryside in classical times and were mostly cut down to build ships and clear land for farming; and the carob trees, the source of the pods used for a chocolate substitute, which flowers in October, yielding a rare honey that comes at the end of the bees' foraging season.

Sicily, as a whole, is famously fertile. Homer told of how Odysseus, returning from Troy, sailed around such an island, marveling at its golden fields of wheat. It became the bread basket of the Roman world and was known, also, as the island of many fruits. Orchards still flow up the sides of Mount Etna, probably the mythical home of the Cyclops (the crater may have been the single eye of the giants). The volcano can still erupt with fury, but between such devastation, the local farmers have benefited from the rich volcanic soil.

Sicily's bountiful agriculture is the base of the island's celebrated cuisine that is layered, like its architecture, with the cultures that have formed this epicenter of Mediterranean civilization: Greek, Roman, Arab, Norman, Spanish, French, and Italian. All these cultures brought their plants and needs; all of these had an effect on the honeys and their uses in the kitchen. This was to be the next part of my search of the classical world.

AN HOUR'S DRIVE down the hillside from Sortino is Syracuse, one of the foremost cities of the ancient Greeks. Syracuse was much associated with the rise of cooking as an art; the first cookbook of the Western world is said to be by Mithaecus, written here in the fifth century BC. In the fourth century BC, Socrates spoke of the refinements of Sicilian cooking, and the fame of the tables of Syracuse, in particular. The city was home to the first school for professional cooks and through this became linked with food, far and wide. It was something of a status symbol among high-rolling Romans to have your kitchen run by such a Sicilian, and he became a stock character in comedies.

One of the best records we have of ancient Greek food comes from Archestratus, a Sicilian gourmet who traveled around the Greek world, recording the gastronomic highlights of more than fifty ports. The sixty-two fragments remaining have the tone of the enthusiast urging his fellow travelers toward the best. If you find the flat-cakes of Athens, *do* try them with the Attic honey, he writes.

Despite his joking tone, Archestratus's words give us some clues about ancient tastes, and help us to see how honey was used in Greek cooking. It was one of the two main sweeteners of the day, alongside boiled-down grape juice. Such sweetness worked as a counterpoint to what the British chef and classics scholar Shaun Hill has identified as "a rank, slightly rotting quality" prevalent in the food of the time; a juxtaposition that was not unlike the contrast between our Stilton and port or mutton and red currant jelly.

Honey was later much used in the somewhat overblown dishes—dormouse in honey and the like—described by the Roman writer on food Apicius in the fourth century AD. He put it in nearly all his sauces. But Greek cooking, as recorded by Archestratus, had a much more refined air. His maxim, that you should use the best, seasonal ingredients and not mess around with them too much, is in tune with many of the best chefs today; these master craftsmen—in the ancient world or the twenty-first century—dare to let their ingredients sing their own flavor notes with an unfussy clarity.

The tables of Syracuse were still exceptional. I'd read of ancient Greek cheesecakes flavored with honey. At Jonico a Rutta è Ciauli, a restaurant above the sea edging the city, I was offered a simplified version of the honeyed cheesecake: a starter of softly scented orange-blossom honey from Sortino with aged pecorino and *cacio cavallo*, a hard cow's-milk cheese from Ragusa, on the other side of the Iblean mountains from Sortino. I took a teaspoon of honey and zigzagged it over the cheese, its gleam turning light to syrup. The matte texture and tangy flavor of the aged cheeses counterbalanced the smooth sweetness of the honey.

As I sat at the table, eating and drinking, I noticed how everything on the table—the cheeses, honey and bread, the white wine from Marsala—was a form of gold. The wine and the honey held the same hue, the cheeses a paler shade, the bread's crust a darker one. Each one of these foods—not least the honey—was virtually unchanged from ancient times. Vincent Cronin's quest for a golden honeycomb ended with wild honey; mine at this simple feast in a hospitable restaurant in Syracuse. It was here that I finally understood how honey had been eaten in the same way for millennia; and how centuries could dissolve and yet form a whole. "The past is not over," William Faulkner said. "In fact, it's not even past."

CHAPTER FIVE

CANDLELIGHT AND INTOXICATION

In the woods that once covered swathes of northern Europe, men would raid the bees' nests they found within hollows in tree trunks. To the bee, a tree is not just a home; the flowers offer an enormous stash of nectar gathered together in one place. Even a single lime or chestnut, say, can release enough nectar to make at least a kilo of honey in just a short amount of time.

The insects' woodland stores then made rich pickings for medieval honey hunters. Russian bee men, for example, working the forests around Moscow, would keep an eye out for swarms in the springtime and follow them to where they settled. Sometimes they smeared honey in a box to trap a single bee, and, on its release, followed the insect back to its colony. A bee man claimed a found nest by carving his personal mark—perhaps a hare's or goat's ear—on the bark, and returned year after year to see if it contained any honey.

Wild forest honey-hunting began to give way to more organized practices. In addition to plundering natural nests, the specialist woodland hunters gouged holes to create nest spaces in which homeless bees could settle, and would make small doorways over the hollows in the trunks so they could check and collect the honey stores more readily. They also hung hollow logs in the trees, high enough for swarms to settle out of reach of the animals on the ground. One illustration shows a cruel trap, designed to get rid of one such major competitor, the bear: the climbing animal is trapped on a platform placed near the nest and shot at by archers so it falls onto spikes on the ground below.

It might be a 12-mile round-trip to check a "bee-walk" of colonies; the more prosperous bee men went by horseback, the less so by foot. In a forest of five hundred known tree cavities, as few as ten might contain nests. Collecting honey could be as treacherous as unpredictable: perhaps one in a thousand hunters died, not to mention the number of broken limbs that came from falling out of trees when pursued by angry insects.

Accordingly, these woodland workers began to devise safer methods. Hollow sections of trunk were taken from the tree and put on the ground to make primitive wooden hives. In Poland and

German woodland beekeepers removing combs from trees.

eastern Germany, ground-based log hives became folk art and were carved into figures such as bears or human beings, perhaps, in the latter case, with the bee entrance positioned below a man's belt so a stream of bees flew comically in and out of his pants.

Although these bee men laid claim to the honey from their nest sites and hives, it was the landowner who let them walk through the woods to collect it, and he demanded a portion of the honey. If the land was sold, this right of honey payments was transferred with it. The woodland laws of medieval Europe were strict and comprehensive. The Ancient Laws of Ireland, codified by St. Patrick in AD 400, covered wild woodland bees as well as those kept in gardens, and imposed a fine for stealing them, the tariff being specified as "a man's full meal of honey." Neighboring properties also had rights of recompense for "the damage bees did to fruit and flowers"; the benefits of pollination were still unknown. As to what constituted the neighborhood of a hive, it was said that a bee flew as far as the sound of a church bell, or a cock's crow, an evocative definition of a local food.

In ninth-century Wessex, King Alfred decreed that a swarm should be announced and claimed by banging metal, neatly turning into an audible declaration of ownership the classical belief, known as tanging, that swarming bees could be made to settle with the sound of clashing metal. England's Charter of the Forests in 1225 established that taking someone else's honey and beeswax was an act of poaching. The judgments of special courts created to enforce forest laws have left us some of the names and misdemeanors of these men-of-the-woods. In 1299, several men were caught at Ralph de Caton's house with a nest of wild bees, leaving behind the remnants of the tree they had burned in order to collect the comb. They were fined for both arson and looting. In 1334, another two men were fined for carrying honey out of Sherwood Forest. But in 1335, a court upheld Gilbert Ayton's defense that he was entitled to the 2 gallons of honey and 2 pounds of wax in his possession; they came from his own woods, and therefore belonged to him.

These rights of honey ownership continued to be exercised in the British Isles long after the medieval period. Even as late as 1852, a landowner in Hampshire's New Forest laid claim, in court, to any honey found in his woods. When it came to wild food, finders were not necessarily keepers. The quantity of laws that relate to honey and wax shows their economic importance. In Germany until the seventeenth century, the chief value of the forests was not just hunting but also harvesting the honey and wax.

The great woodlands of medieval Northern Europe progressively dwindled, as trees were felled to build ships or burned to smelt metal. Woodland beekeeping continued as long as there was honey to be collected; meanwhile, another way of keeping bees was being developed in northwest Europe where there were fewer large trees to provide material for log hives.

THE FIRST of these new hives in the British Isles were made from wicker woven into conical shapes and covered with a substance called cloom, which was cow dung mixed with something like lime, gravel, or sand. These structures sound like an agricultural version of the wattle and daub of human habitations. They were mostly replaced by the skep, the domed straw hive that has become a much-loved symbol of traditional beekeeping. The word may come from the old Norse *skeppa*, meaning a basket that both contained and measured grain. Skeps were originally used by Germanic tribes west of the Elbe and entered the British Isles through East Anglia with the Anglo-Saxons; the hives also spread southward as far as the Alpes-Maritimes and Pyrenees in France. Today, these rustic hives are made pretty much in the same way as they were in medieval Europe. They still have their fans among beekeepers because they insulate the bees better than wood, and their rounded shape suits the clustered ball of overwintering bees, helping them survive the cold.

The materials for skeps came originally from the agricultural materials that were at hand. Lengths of wheat straw were used, or dried stalks from other plants such as reeds, rye, and oats. These were gathered together and pushed through a section of cow's horn with the tip cut off to create an aperture about $1\frac{1}{2}$ inches in diameter. (A twentieth-century skep maker, Frank Alston, suggested substituting the cow's horn with the metal ring of a TV aerial.) The gathered bunch of straw was then coiled into a dome shape and bound together with a strong, flexible material, which was traditionally dethorned bramble briars, though cane is now used, more conveniently.

Some skeps were given straw hats, known in parts of the country as hackles, to keep out the rain; others were sited in shelters. One elaborate medieval shelter in Gloucestershire, made of Caen stone, has niches for thirty-eight skeps and magnificent carved dividing brackets between them. It was in danger of demolition until given a resting place, first at the agricultural college at Hartpury, and now at the local church.

On a more domestic scale, there was the bee bole, a recess into a house or garden wall in which one or more skeps were sheltered from the elements. These tend to be sited on north- and east-facing walls to protect the bees from the prevailing southwesterlies. You can see them still in the cob walls—some of which are as much as $4\frac{1}{2}$ feet thick—of traditional Devon buildings as well as in other rained-upon, traditional places such as the Lake District and the Yorkshire dales. One Cumbrian bee bole can be spied in an illustration in Beatrix Potter's *Tale of Jemima Puddleduck*. Beatrix Potter took her images from her surroundings and in real life you can see this very bee bole beside the vegetable patch at her house Hill Top, near Ambleside. In mid-Devon there are two, high up in a wall of the Ring O'Bells pub at Cheriton Fitzpaine, a village near Exeter. Bee boles come in different shapes in different parts of the country: in Scotland they are rectangular; in Devon, arched or domed; and in Kent, gabled or triangular. These skep niches are an example of the local distinctiveness that has been so eroded in this homogenized world, but is still there to be enjoyed in many rich particulars, once you are on the lookout.

KEEPING BEES in skeps was different from modern beekeeping in one vital aspect: in order to cut out the comb, many people destroyed the colony. The skep was put over a pit that held a burning paper dipped in brimstone; this gassed the bees with sulfur. A cabbage or rhubarb leaf might be put on top of the smouldering fire to stop it from being extinguished by the falling, dead bees—a curiously homey detail for a deadly task. Care had to be taken not to adulterate the comb, and the sellers could be taken to court if this happened.

Beekeepers tended to remove the comb from the heaviest and the lightest skeps; in the case of the former, because it would contain the most honey; and in the latter, because the bees probably wouldn't survive the winter on the honey within in any case. There was much disquiet about the killing of these sacred and useful creatures, and people devised ways of driving them from one skep into another so the comb could be taken from an empty nest. One of the disadvantages of this annual slaughter was that it bred out the more productive strains. Nevertheless, the colonies left to survive the winter would, if all had gone well, get going again in the spring, when they would reproduce and swarm once the skep was too small. The breakaway bees would be caught in new skeps, to start all over again. Straw skeps are used by some beekeepers today precisely to catch swarms; their light, handy shape is good for this task.

I went to see a contemporary skep maker and beekeeper, David Chubb, who lives in south Gloucestershire. His farmhouse was in South Cerney, a village that has been colonized by the commuters; there were plenty of four-wheel-drives on the roads, raring to tackle the wild, urban terrain of Cirencester and Swindon. In a previous life, before he went into farming and skep-making, David was a maintenance mechanic on the railways. One of his best honey crops comes from the wildflowers on the disused Southampton-to-Birmingham railway line. It was incongruous but somehow satisfying that the old-fashioned skep was being made in a place that so defied the picture postcard.

The Chubbs have moved through mixed farming, diversifying long before it was the trend, to keep the big, rare-breed Cotswold sheep, chickens for eggs, and bees, as well as doing contract work and making skeps. David sometimes puts his Cotswolds onto nearby land to act as bucolic lawn mowers, and problems can arise when people want to fit the sheep in like appointments; when commuters live alongside farmers, neither quite speaks the other's language. But this is a modern village, where tight estates of new houses face onto fields, and both tribes need to talk to each other if the incomers are to get their rural dream and the farmer is to survive. Both have an attachment to the idea of the countryside, as well as its reality. When I asked David what he liked about the countryside, his blue eyes became momentarily abstracted. "I escaped over the garden wall when I was four," he said, "and I've never gone back. It's better out there. Nature."

What better way to be a part of your place than to eat your surroundings in the form of honey? David now focuses mostly on the skeps and the bees, selling his pots in local shops and from his home.

He started making skeps because he couldn't get hold of one himself. Some people like them for display because the golden dome
represents a certain image of rusticity; like the handicrafts made for
the heritage pub market, the symbol lingers beyond use.

But at least half of David's customers still use skeps for beekeeping, often buying them from him via the modern means of the
Internet. At the Chubbs' farmhouse, we looked at photographs of
Dutchmen with skepfuls of bees at markets—the tradition of these
hives continued longer in the Netherlands—and at pictures of
skeps that showed how the bees' entrance hole could be found at
different places, at the bottom or the top, and at images of another
kind that looked like a medieval helmet for a scarecrow, with a single eye-slit halfway up. David agreed that skeps provide good insulation (he said an inch of straw was equivalent to 6 inches of wood),
and he continues to keep at least one colony in this way as an insurance policy against a big winter freeze. Mostly, though, he keeps his
own bees in wooden hives, because it is so much easier to harvest
the honey from them.

I watched David in his skep workshop. He buys old-fashioned
wheat from the same sources as thatchers, who also need longer
stalks, harvested with a binder to preserve their length. The wheat
still had its ears on, and David pulled them off as he went along. As
he talked, he tamed the rustling tail of wheat-ears into a fat coil, a
process that reminded me of hairdressing, and I remembered how
the medieval skep's domed outline and coils were later repeated in
the 1950s and '60s beehive hairdo.

We went to David's honey room, where the jars of glowing
honey resembled big, boiled sweets, then moved outside to see his
skep. It looked like a huge, straw thimble; when David turned it
over, I could see the layers of honeycomb in beautiful, organic
swags, massed with bees—a mesmerizing sight.

Driving away from the gold skep and gold honey, and the shorn

Sic nos non nobis mellificamus apes.   Omnia in libris

All plants yeild honey as you see
To the Industrious Chymick Bee

"So we the bees make honey, but not for ourselves": a seventeenth-century woodcut showing a skep.

wheat fields that were Scandinavian blond in the bleaching light of late August, I thought again of how the skep was still the single most evocative image of beekeeping, instantly conjuring up a cottage garden full of drowsily buzzing bees in late-afternoon sun—even if, in reality, it was now made in a commuter village and sold to customers who did not use it for beekeeping at all.

DEEP IN THE CONFUSING forests of Shakespeare's *A Midsummer Night's Dream,* Titania ordered her fairies to find wax candles to light the way for her newly beloved Bottom (act 3, scene 1):

"The honey-bags steal from the humble-bees, / And for night-tapers crop their waxen thighs, / And light them at the fiery glow-worm's eyes, / To have my love to bed and to arise."

Shakespeare shows the contemporary confusion between the pollen loads on the bees' legs and the wax they secrete from within themselves.

One of the most extraordinary skills of the honeybee is its production and use of beeswax. When the bees have no more comb in which to store honey, the nectar they have collected stays in their honey stomachs and the stored sugars assimilate into their bodies; this enables them to secrete small, transparent plates of wax through the glands on the front of the abdomen. The wax is moved by their legs to their mouth parts, where it is kneaded until it is ready for use. With this new wax, they build more comb.

The honeycomb is built by a hanging curtain of worker bees, who pass the materials up a chain to their fellow builders. The cells of the honeycomb are hexagonal because this design needs the minimum amount of wax to hold the maximum amount of honey—almost $2\frac{1}{4}$ pounds of honey in just 1 square foot of comb. The bees are so careful with their wax because it takes nearly 16 pounds of honey to produce enough comb for a colony: wax secretion is therefore a very expensive item in the hive's energy economy.

Light as a feather, the wax comb is now ready for use. It hangs in parallel, vertical sheets, a network of regular cells placed back-to-back and tipped on a slight angle so the nectar and honey do not drip out.

～

HONEY WAS NOT always the most valuable product of the bees; in medieval Europe, beeswax was so important that it could almost be said that honey was a by-product of the wax. In England, at the time, the wax could be worth eight times as much as honey. No wonder, when it provided the most prized means of lighting after dark. Just as the bees can manipulate the warm wax to make comb, humans can easily mold and shape molten wax. It is flammable, as well as malleable, and so can be wrapped around a piece of fabric, or a wick, to form a candle.

The cheapest candles of the medieval world were rush lights. Reeds were picked, ideally, in the early summer when they were young and juicy, stripped to just a single layer covering the pith, and dipped in melted fat. These rush lights could be made from beeswax, as Pliny the Elder mentions in the first century AD, but mostly they were made with animal fat, or tallow. Sheep fat was the best, with beef fat as second choice. The pig, a commonly kept animal that could be fed on household scraps, unfortunately had fat that burned with a thick, black smelly and greasy smoke; since pig fat tastes far better than sheep fat, perhaps this was ultimately providential. The smelly, sputtering nature of tallow candles shows why, in contrast, beeswax was so valued. Beeswax candles have a beautiful, clean, and cozy scent of honey, and emit a pure, unsmoky light. The seventeenth-century bee author, Reverend Charles Butler, summarized its merits: ". . . it maketh the most excellent light, fit for the eyes of the most excellent; for cleernesse, sweetness, neatnesse, to be preferred before all other."

There were many kinds of beeswax candles. Perchers were tall candles for altars and ceremonies. *Quarerres* were large, square candles that were used in funerals. Flambeaux—torches made from material soaked in resin and coated with beeswax—were designed to burn so fiercely and brightly that they could be kept flaming in wind and rain and taken on processions. The way the cloth was twisted may have given rise to the word *torch* from the Latin *torquere*, to twist.

In great halls, wax candles of many sizes were set in candelabra. The brightness of their light enhanced the status of the lord of the household, and candles were part of the payments for members of the household. Domestic accounts go into some detail on such allowances, showing just how valuable wax was. In the fifteenth century, the Lord Steward of Edward IV had a winter allowance that included a torch to attend to himself, a *tortayes* (small tree) of candles to set at his livery basin, three perchers, and seven tallow candles.

Beeswax could be readily available, if you kept bees, in which case you could make the candles yourself; but by the late thirteenth century, demand outstripped supply, and much wax was imported, mostly from Europe through the Hanseatic League. There were now also dedicated candle makers. These wax chandlers rose to be one of the eighty-four livery companies of the City of London. They started as a "mistery," which may come from the Latin *ministerium*, meaning occupation (the French word *métier* comes from the same root). Misteries were composed of master craftsmen, who served an apprenticeship and formed an organization that represented the skills and pride of a trade, maintaining standards by such means as fining makers who sold adulterated candles. In 1482, the wax chandlers were honored with a charter, decorated on its borders with honeybees amid flowers and bearing the royal seal made of resin mixed with beeswax. With the honor of a charter, like the other guilds, the wax chandlers were "invited" to contribute money to the Crown to swell its war coffers.

The chandlers made candles in various ways. Molten wax might be dripped down a wick, building up in successive layers; a wick might be drawn through melted wax; the soft wax could be rolled around a wick; or it could be put in a mold with the wick placed in the middle. Some candles were really just long lengths of wick barely covered in wax. Long coils of these thin tapers were called trundles; in fourteenth- and fifteenth-century France, lengths of fiber, long enough, it was said, to encircle a town, might be dipped in wax and burnt as a protective charm. Whether or not one literally did go around a town, the idea illustrates how candles were both practical and symbolic. Nowhere was this more evident than in the Christian church.

~

THE CHURCH was a voracious consumer of wax candles in the Middle Ages. Monks kept bees, and rents and tithes from their substantial lands could be paid by their tenants in wax. Candles were needed in large quantities to light the large, dark spaces of churches, chapels, and cathedrals. In the eighth century, Pope Adrian I burnt a perpetual light in St. Peter's in Rome consisting of 1,370 lights in the shape of a cross: it must have been a staggering sight in the dim world of the Dark Ages.

Candles have a many-layered significance in Christian belief; not only was light seen as symbolic of the awakened soul, but beeswax was regarded as pure because of the chasteness of the worker bees. If the wax was the spotless body, the wick was the soul and its flame a symbol of divinity, or the Holy Spirit. Some took the metaphor further, to portray the drones as monks, and their autumnal expulsion from the hive as a symbol of good moral housekeeping by the punishment of the lazy. Pure beeswax candles were burned at liturgical services, and there was a belief that the souls of the dead could only be at peace when watched over by the living. Candles were burned in their memory, a custom that continues today in the slightly different form of votive candles.

Honey, too, was also part of early Christian custom. Until the seventh century, people took honey and milk just after being baptized, a rite that echoes the early diet of the infant Zeus. In another story that connects the medieval world with that of the Greeks and Romans, St. Ambrose, the fourth-century Bishop of Milan—and the patron saint of beekeepers—was visited by bees in his cradle. The insects flew up high, vanishing as if to heaven. This was said to be a sign that St. Ambrose would be both great and eloquent.

Candlemas on February 2 is the day when the candles for the year are blessed and distributed. Based around the same time as the

Roman pagan festival of purification, it is the festival of the purification of the Virgin Mary and marks the presentation of the infant
Jesus at the temple, when St. Simeon, in the Nunc Dimittis, called
Jesus "a light to lighten the gentiles." By the middle of the fifth century, candles became an explicit feature of the occasion.

The guilds, including the Wax Chandlers, would participate in
big religious festivals, and doubtless provided the candles. The
famous funeral procession of Henry V in 1422, an exercise in state
showmanship of some proportions, was lit by fourteen hundred wax
tapers along the 2-mile procession, and every fifth man held a torch.

The significant religious demand for beeswax meant it was dealt
a great blow when the monasteries were closed by Henry VIII in the
sixteenth century. Although there was a revival of candle use under
the Roman Catholic Mary I, exemplified by the making of an enormous, 300-pound Paschal candle, such displays fell from favor in
England. Oliver Cromwell later banned altar candles altogether.

Beeswax candles were an especially valued form of light; then in
the eighteenth century, New England whalers discovered how to
burn the oil found in the head of sperm whales, in a substance
called spermaceti, which was refined into a hard, crystalline substance and made into candles that burned with a new brightness.
The cheap, mass-produced paraffin candles of the nineteenth century were to further eclipse the light of the honeybee. You can,
however, still find beeswax candles as a specialist product and enjoy
their soft scent and beautiful shine. The Wax Chandlers Guild continues and has links with contemporary beekeepers, offering an
award if an entrant especially excels in the British Beekeepers
Association annual examinations.

∽

IN MANY CULTURES around the world, honey was regarded as
a magical, transforming substance, almost a potion. At the heart of

this belief was mead. As a sweet intoxicating liquor, it was the main use of honey—more so than as food or medicine. The question arises, even, whether honey was more valued for its sweet taste or for its power as an alcoholic drink.

The sugars in honey naturally ferment when mixed with water and yeasts already present, or in the air; it is therefore one of our most ancient drinks. The Greeks had many honey-based drinks, some of them, like mead, made from fermented honey and some made from honey mixed with another drink, such as the common mulsum, a mixture of wine and honey. Rhodomel was made with roses and honey; omphacomel from grape juice and honey; thalassiomel used seawater; and there was even a meal-of-a-drink called kykeon, with oil, wine, cheese, and mead, that was drunk at the harvest feast. Dionysus, or the Roman Bacchus, was probably the god of mead as well as wine. But honey drinks were also a source of good health and not just bacchanalian excess. Pollio Romulus, aged one hundred, told Julius Caesar he had kept the vigor of mind

The impact of mead on some Caucasian peasants.

and body by taking spiced mead inwardly and using oil on his outer body, an alcoholic variant on Democritus's prescription for long life: taking honey on the inside and oil on the outside.

Mead predates wine, perhaps by many thousands of years; but when the grape arrived, it eventually began to replace mead in southern countries. Mead continued as a fine tradition in northern climes. Pliny wrote of the British Celts that "these islanders consume great quantities of honey-brew." Mead was a drink of status, of royal courts, actual and literary. The Queen bears the mead that is given to Beowulf, and King Arthur downed gobletfuls of the stuff. Mead is much mentioned in Welsh poetry and laws. A free township had to pay its dues to the king by giving him a vat of mead big enough for the monarch and a companion to bathe in.

Great ceremonial communal drinking bowls were made of wood and precious metals, and known as mazers, from the word for one of the original materials, maple. One example of such a vessel is the Rochester Mazer, made in 1532, which is in the British Museum in London: the wide, shallow bowl would have been held with two hands and passed between drinkers. (Admiring this renowned object, now sequestered in its glass museum case, I allowed myself the scurrilous thought that there might have been a term, in the ebullient vein of northern European drinking traditions, "to get mazered.")

Beyond the British Isles, mead was the drink of gods and heroes. In Teutonic mythology, warriors reaching Valhalla quaffed sparkling mead offered by divine maidens. Odin stole the magic mead of the gods, made love to its guardian giant's daughter, drank the brew in three mouthfuls, and fled in the form of an eagle.

Some of the intoxicating stories connected to mead have come down to us, almost inevitably, with a lurching exaggeration. In the tenth century, Olga, widow of the Prince of Kiev, was said to have invited the murderous mourners to her husband's funeral, instruct-

ing them to bring plenty of mead. These five thousand incapaci-
tated drinkers were then slain in vengeance. Meissen, in what is
now southeast Germany, had so many breweries that mead was
supposedly used instead of water to put out a fire in 1015. In 1489,
Tartars attacking the Russians discovered a stash of Russian
mead—perhaps put there as a deliberate trap—and drank the
booty, to be easily overcome.

Mead was also an aphrodisiac that enhanced love, smoothed its
way and powered its virility. The Scottish saying that mead
drinkers had the strength of meat eaters probably did not refer to
the bulk of their biceps. The alcoholic properties of this honey
drink may well have enhanced honey's reputation as a love food.
The term *honeymoon* may originally come from the initial, sweet
bout of postmarital passion—though there is another theory that it
stems from the relatives' monthlong mead drinking in celebration
of the nuptials.

The remnants of mead that you find now tend to be sweet, not
unlike a fruit wine. Mead in its heyday had many strengths and flavors. It could be brewed strong and dry, distilled, or made to
sparkle. It could be a weaker, quaffing drink or flavored with herbs
and spices, in a drink that became known as metheglin. This aro-
matic, medium-strength drink sounds like an ancestor of today's
grape-based aperitifs, such as vermouth.

Sir Kenelm Digby (1603–1665) collected recipes that were published after his death in *The Closet of the Eminently Learned Sir
Kenelme Digbie Kt, Opened*; he had more than a hundred recipes for
mead and metheglin. The drinks he described could be kept for a
few months or for up to three years. One reason such importance
was placed on alcoholic drinks at that time was because they were
a safe, as well as delicious, alternative to polluted water. Sir
Kenelme Digbie's championing of mead also had a background of
disquiet about imports. The deluge of wine coming into the coun-

try from the rest of Europe made native drinks seem out-of-date. Mead was made simply with honey, water, and yeast and was a good, honest, home-produced brew.

The romance of mead, with all its sophisticated variants, did not last. Today the drink seems faintly quaint and can be a touch cloying due to the preponderance of the sweeter meads, which capitalize on the honey flavor, rather than the drier styles of this versatile drink. There are some delicious brews still made commercially (as well as the home brews made by curious beekeepers who can spare the 4 pounds of honey it takes to make a gallon of mead). When well made, mead can give an initial impression of a sweet, honeyed roundness that turns into a mellow dryness similar to a fine sherry.

The superseding of mead by other drinks came with the rise in the price of honey. Sugar became more available (though it was still luxuriantly expensive until the advent of the sugar beet) and the Reformation meant fewer church candles and therefore fewer bees. But although the number of hives declined, the bee did not drop in status. If anything, these sacred insects were to become even greater objects of fascination in the scientific age that was to come.

CHAPTER SIX

ENLIGHTENMENT

The Reverend Charles Butler signaled a new phase in man's understanding of the honeybee. He used his own observations to challenge medieval beliefs, therefore pushing back centuries of received wisdom. An era in which scientific thought became based on the direct study of the natural world was about to begin.

Butler was the vicar of Wootton St. Lawrence, near Basingstoke in Hampshire, in the early 1600s. As well as being a scholar, teacher, and inventor of a form of phonetic spelling, he wrote about logic, music, and theology, was an advocate for the legality of marriage between cousins, and was the great-great-grandfather of the eighteenth-century Hampshire naturalist Gilbert White. He is best known today as one in a line of humane, engagingly helpful authors on the subject of the honeybee; his book *The Feminine Monarchie* is still read by beekeepers with affection.

The Feminine Monarchie, first published in 1609, was well received at court, and the third edition (1634) was even dedicated to Queen Henrietta Maria, the start of a fashion for presenting such books to the royal "queen bee." Butler starts with a note of admiration for the bee's moral rightness and practicality, depicting the relationship between bee and beekeeper as one of good, clean living. If the beekeeper approaches the hive drunk, puffing, blowing, hasty, and violent, he will be stung; if he behaves properly, all will be well: "[T]hou must be chaste, cleanly, sweet, sober, quiet and familiar; so will they love thee, and know thee from all others."

The righteous praise for such steadfast qualities is leavened by Butler's varied and idiosyncratic enthusiasms for his insects. The sound of a hive starting to swarm is well known to beekeepers. Butler, a music scholar, scored it in the first edition, later turning the notes into a madrigal with four parts. Another section of his book deals with such matters as sealing wax, drinks, and a syrup of violets and honey that "tempereth and purgeth hot and sharp humours, expels melancholy and effects—headache, waking, dreams, heaviness of heart," like a seventeenth-century liquid form of aromatherapy. It was Butler who recorded the flavorings of Queen Elizabeth I's favorite spiced mead: rosemary, bay leaves, sweet briar, and thyme.

Charles Butler's most significant advance was to trust his own eyes on the workings of the hive, rather than the words of classical authors. He notes, for example, that drones were clearly part of the reproductive means of the hive, although exactly how, he did not know. He also observes the "shivering" bees that seemed to presage swarming. We now know this is one way bees communicate within the hive.

Much of the book is immediately recognizable. Nature, after all, barely changes, just our knowledge of it. Butler describes the poisoned spears of stings that you must quickly flick out to stop the pain being "greater and longer"; the late swarms, known as blackberry swarms, which start their new colony too late to store enough honey to survive the winter; the passing of the year through the zodiac and its different flowers. Gemini brings us honeysuckle and bean blossom. Cancer brings the thyme that "yieldeth most and best honie." Then come knapweed and blackberry, and ivy during

Scorpio. In this classic of bee books, you walk through the past as if it were your back garden.

~

BUTLER'S WORK was just the first break from classical tradition; in the seventeenth century, new technology would start to unravel many mysteries of the honeybee. People could now examine bees more closely in two ways: through observation hives, which opened up the internal workings of this city of insects, and through microscopes, which revealed the individual bee in minute detail.

One of the first observation hives in England was made by the Reverend William Mewe, rector of the parish of Eastington in Gloucestershire between 1635 and 1655. Although Mewe considered becoming a polemical writer, and took the republican side during the civil war, he ultimately retreated to the countryside, where he mused on such subjects as the honeybee.

Mewe's interest in bees was partly moral. His hive was inscribed in Latin, praising the insects' industriousness and harmonious community. For him, the bee's well-ordered colony was an example to mankind. Civil war and regicide had torn England in two; the country needed to find a cohesive social order, and what better example than the productive government of the hive? "When I saw God make good his Threat, and break the Reines of Government, I observed, that this pretty Bird [the bee] was true to that Government, wherein God and nature had set it to serve," wrote Mewe. As well as being able to look at the bees in their glass-paneled hive for moral instruction, he thought the bees also benefited: they produced more honey, he believed, from the very fact of being watched.

Mewe was a rural cleric who kept his observation hive in his garden; all the same, his ideas soon spread to the world beyond.

Word of the hive may have reached Oxford through Samuel, one of his eight children, but it is just as likely that the design was sought out by others who took up the cause of beekeeping in the 1640s and 1650s.

The first known description of a windowed hive comes in John Evelyn's diary entry for July 13, 1654. He records dining with "that most obliging and universally Curious" Dr. Wilkins, at Wadham College, Oxford, where he saw transparent apiaries built like castles and palaces, adorned with dials, statues, and weather vanes. King Charles II later came especially to see these hives, which he contemplated "with much satisfaction." The design of this hive came from Mewe.

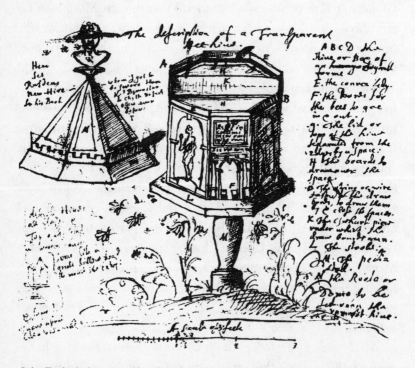

John Evelyn's drawing of his observation hive, with adornments, given to him by Dr. John Wilkins.

The "universally Curious" Dr. John Wilkins (1614–1672) was at the very center of seventeenth-century intellectual life. Whether as Warden of Wadham College, Oxford, as Bishop of Chester, or as joint first secretary of the prestigious new body for scientific discovery, the Royal Society, he was consistently in the midst of matters religious, scientific, and political, and deftly managed to remain so both during the Commonwealth—when he married Oliver Cromwell's sister, Robina—and under Charles the Second. He was also one of the most significant beekeepers of the century.

Wilkins's gallery and lodgings at Wadham, as described by Evelyn, sound rather like his hive, full of "dyals, perspectives . . . and many other artificial, mathematical, Magical curiosities; a Way-Wiser, a Thermometer, a monstrous Magnet, a conic and other Sections, a balance on a demi Circle, most of them his owne and that prodigious young Scholar, Mr Chr[istopher] Wren." The rooms embody the mind of its polymath inhabitant; Wilkins's conversations and thoughts, his dreams and inquiries, can all be glimpsed in this intellectual toolshed. His projects included a sail-powered coach, a double-barreled wind gun, and a machine for weaving ribbons. He was part of a group whose conversations and experiments cross-pollinated and bore fruit. Christopher Wren came upon the idea of injecting animals with liquids after discussions with Wilkins and Robert Boyle, and these were later to become the famous experiments in blood transfusions carried out by the Royal Society.

The breadth of inquiry within this circle included a fascination with the honeybee. One of their quests was to collect honey without killing the bees. Straw skeps began to have extensions put on top, where the honey could collect and be removed without destroying the rest of the hive. These resembled a primitive version of the wooden box "supers" used by beekeepers today.

In the seventeenth century, the search for the perfect hive mirrored the contemporary preoccupations with science and architecture. The hive in Wilkins's garden, with its vanes and dials, displayed the abiding interest in meteorology. A drawing of one such hive, owned by John Evelyn, shows statues and other ornamentation, but its essential structure was octagonal, which was the cabinetmaker's approximation of the round nests of the honeybee. Stripped down, as it was in one of Christopher Wren's first architectural drawings, Evelyn's hive was a stack of boxes on top of one another, with holes between them so the bees could leave their honey in one box and crawl down to the next; the honey on top could then be removed—or at least this was the theory.

OF ALL THE seventeenth-century devotees of the honeybee, perhaps the most significant was the writer and collator Samuel Hartlib. Hartlib collected thoughts and experiences from an impressive network of contacts from England, Europe, and the New World with the purpose of bringing stability, prosperity, and healing to England through the scientific advance of agriculture and horticulture—motivated, also, by the starvation in the country following the disastrous harvests of the 1640s.

Of the twelve books Hartlib published on food production, *The Reformed Commonwealth of Bees* (1655) was typical in bringing together ideas and experiences from a range of sources. In the book, he published a number of letters on the insect, including those from William Mewe and Christopher Wren. Hartlib also exchanged scientific views with Dr. Wilkins from 1649 on, and the first recorded use of the English word *apiary* was by John Evelyn writing to Hartlib in 1650.

The Reformed Commonwealth of Bees shows a belief in the economic possibilities of the honeybee. An estimated $177,500 worth of honey could be made if hives were put into every parish of the land. Eyes were cast toward France, where fine Bordeaux honey fetched more than ten times the price English honey could command. The sugar plantations in Barbados were becoming economically significant, but Hartlib showed a bias toward English produce; he also pondered the possibility of extracting sweetness from home-grown apples.

So much for the theory. This group of seventeenth-century intellectuals did have practical beekeeping experience—for example, John Evelyn recommended in *Sylva*, published in 1664, particular trees that were good for bees, such as the oak, the black cherry, the poplar, the willow, and the buckthorn with its "honey-breathing blossom"—but how successful, ultimately, were their hives? These new boxed hives were not universally popular. One user wrote to Hartlib in 1658 saying it made no difference to the amount of honey collected, and that plain country traditions such as skep beekeeping produced more profit for less trouble. Perhaps these enlightened beekeepers set too much store by the rationality of the bee. Instead of crawling into the hive and going up to the top and building downward, the bees started their comb in the bottom box. The holes between the boxes did not allow a large number of insects to move easily into the upper layers. Furthermore, the elaborate, hexagonal wooden hive was too expensive and difficult to construct to be economical for commercial beekeepers.

For all their thoughts, inventions, and curiosity, these enlightened Englishmen were only at the beginning of the discovery of the bee. It was another two centuries before the problem that preoccupied them—how to remove the honey cleanly, without killing the bees—was finally resolved. Regarding the scientific study of bees, the early observation hives had just small panels of glass, offering only a partial view of the colony. Other scientists were to take matters much further.

~

ON THE CONTINENT, a young Dutch scientist devoted months to focusing entirely on the honeybee. By using observation hives and peering through the glass eye of a microscope, he went further than anyone else had in understanding the insect's mysteries. Jan Swammerdam was born in 1637. By the time he was thirty-two, he had written his *Historia Generalis Insectorum*. It was one of three books published within two years that marked the beginnings of entomology—the others were Marcello Malpighi's study of the silkworm and Francesco Redi's book on insects. The young Dutch scientist went on to become one of the pioneering students of the honeybee.

Swammerdam grew up in Amsterdam. His father was a noted collector, who kept a renowned cabinet of curiosities on display at his home. New objects, from Chinese porcelain to fossils, would arrive through the city's port, at the center of world trade. Visitors also came on the ships, making their way to the house to study and admire the collection.

The young Swammerdam began to create a natural history collection, picking up insects and their eggs, food, and even their excrement on expeditions both in Amsterdam and in the towns and countryside beyond. He searched the air, land, water, meadows, cornfields, sand dunes, rivers, wells, trees, caves, ruins, and even privies in order to find his quarry. His findings were also put on display; by the age of twenty-four he had no fewer than twelve hundred items to show, a number that would eventually more than double.

There were now two collectors, of different generations, in the same family, sharing the same house. This must have created tensions. The relationship between father and son was to prove difficult on many levels. Conflict manifested itself, at first, in Jan Swammerdam's choice of profession. His father, a pharmacist, wanted his son to go into the church. Jan Swammerdam, although a deeply pious man, did not feel his temperament would suit a ministry. Following his own passionate interest in nature, he decided to become a doctor instead, and went to study anatomy, surgery, and medicine at the celebrated University of Leyden.

The microscope was a new tool of discovery in the seventeenth century and was first used to make magnified drawings of the honeybee in 1625. These bees were presented as illustrations of the Barberini family crest, rather than specifically scientific diagrams. Maffeo Barberini was Pope Urban VIII, and in a powerful position to influence the conflict between questioning scientists and the Roman Catholic Church. The drawings, viewing the honeybee from above, below, and the side, were therefore meant to flatter, and an engraving of the bees was presented to the pope at Christmas as a "token of everlasting devotion." The magnified bees were, however, next published in a literary work—a book of satires by Perseus—rather than an explicitly scientific book. Later in the century, Robert Hooke's *Micrographia* (1665), famous for its illustration of a flea, included some drawings by Christopher Wren and detailed images of the honeybee's sting.

Jan Swammerdam, having studied human anatomy, turned his attention to that of insects, using a microscope. On a visit to France, he met Melchesedec Thévenot, a wealthy French gentleman and diplomat who traveled to pursue his interest in other countries and science. When Swammerdam came to stay at Thévenot's estate near Paris, servants went out to the Seine to collect insects for the young Dutch guest. It was Thévenot who invited Swammerdam to participate in the august gathering of the new Académie Royale des Sciences. Swammerdam didn't say much at this meeting of scientific minds, but contributed by dissecting insects to show their entrails. Thévenot later accompanied Prince Cosimo de Medici to Amsterdam to see the cabinets of both father and son. On this occasion, Swammerdam cut up a caterpillar to show how the butterfly

The first drawings of bees (1625) based on observations through a microscope.

could emerge from its larval anatomy. The prince offered a large sum to the Dutch scientist if he would bring the collection to his court. Swammerdam, a devout Protestant, declined the Catholic's offer.

Swammerdam was by now deeply immersed in his insect explorations. His drawings of the honeybee were made from 1669 to 1673. For months of extraordinary devotion, he would rise with the light, starting at 6 a.m. to work for as long as possible. He worked with the sun beating down, peering at the bees through glass hives and his microscope. By noon, his straining eyes would start to fail. After that, he worked well into the night recording his observations and drawing until he could continue no longer. Despite these labors, Swammerdam still wished he had a year of never-ending light to work harder still. By the end of these studies of bees, the summit of his life's work, Swammerdam's body and mind were battered; some think he never recovered.

Swammerdam's drawings show the exquisite dexterity of his dissections. He used instruments so tiny that they had to be sharpened under the microscope. His favorite tools were tiny scissors with which he could separate out and cut minute parts of the bee's anatomy without tearing them. He would put a bristle into a bee's gut and inflate it by blowing down a tiny glass tube so he could inject a colored fluid and see the anatomical structure more clearly. The insects were punctured with a needle to drain their fluids, dried, and anointed with resin and oil to preserve them. When Swammerdam started his observations of an insect, he would first look at it through a weaker lens and move progressively closer and closer by turning to stronger and stronger lenses. Once he became familiar with the animal's general form, he could put each new detail in context as he perceived it. Using this approach, he was able to discover the ovaries of the queen bee, anatomically proving her role as egg layer.

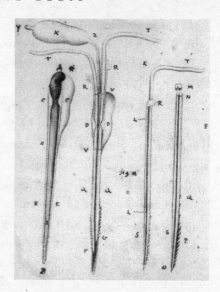

Swammerdam's anatomical drawing
of the honeybee's sting.

By this stage, the pressures of family life were impinging on Swammerdam's devotion to discovery. His father wanted him to earn a living as a doctor rather than wasting his time drawing insects. Financially dependent on his father, Swammerdam had to turn down an offer from his old friend Thévenot to live and study with him in France. His options at home were limited. He retired to the country, but there was surrounded by the insects of his studies. Thoughts swarmed in from every direction. Swammerdam also became plagued by a dilemma of vocation. He saw a conflict between his studies and his love of God. Should he study God alone, or God's works? At times, he had reconciled his studies and religion, writing in a letter to Thévenot: "Sir, I present you the omnipotent finger of God in the anatomy of the louse." But for now he turned away from the natural world and toward its maker.

He tried to sell his natural history collection and began a correspondence with Antoinette Bourignon, a guru known as The Light of the World. She had been left part of the island of Nordstrand, near Schleswig, and lived there surrounded by acolytes. When

Swammerdam wrote to Bourignon for advice about his conflicts between belief and science, her reply allowed him to finish his study on bees but was nonetheless stern. Give up "the amusements of Satan," she wrote, and concentrate on Jesus. Swammerdam left his unpublished bee studies behind, as if carelessly, no longer believing they were the way to God, and traveled to join Mlle. Bourignon's community in September 1675.

The trip was not a success. There were problems both within the group and with the community outside. For a while, Swammerdam acted as secretary and translator for Antoinette Bourignon, but nine months later returned to Amsterdam to a less-than-conciliatory father. Still dependent, the son had to live with the father once again, until the latter gave up his house and went to live with his daughter, leaving Swammerdam even more isolated.

In 1678, Swammerdam's father died. Although his financial problems were eased, his sister took more than her share of their father's estate and his physical condition deteriorated further. By the end of 1679, he was very ill. Thévenot offered to send medicine to help his fever, and Swammerdam asked desperately for a palliative for his edema. But it was too late. The obsessional Dutch scientist died in his forty-third year, destroyed by fever and mental exhaustion: incessant study, anxiety, and illness had worn him out. He died like a worker bee, falling in the field, all energy spent. At a posthumous public auction, the cabinets of both father and son were sold together, though in lots, not as complete collections.

But on his deathbed, Swammerdam returned to his work on the honeybee. Bees displayed God's wisdom and power in a mathematical manner, he said; their minute exquisiteness made the glory of God all the greater. In his will, he asked that his work be published, in Dutch as well as Latin, so it would be more accessible. Even his final resolution was not without complications. Swammerdam had left the ownership of his unpublished engravings of bees to Thévenot; but he sent the engravings themselves to a publisher, who refused to relinquish them until forced to do so by legal action. The drawings were not published until 1737, more than sixty years after they were made, in a book whose title, *Biblia Naturae*, or *Bible of Nature*, combined Swammerdam's devotion to God and his creations. It became his most celebrated work.

Swammerdam's legacy was to pioneer microscopic work on the honeybee, and this helped to sweep away a blind belief in classical learning. In 1880, a plaque was put on Swammerdam's house bearing the words: "His study of nature remains an example for all times." Observational inquiry, rather than the uncritical repetition of erudition, was the scientific path he helped establish.

IN THE EIGHTEENTH CENTURY, observation hives became popularized by the French scientist René Réaumur (1683–1757), who in his *Mémoires pour servir à l'histoire des insectes* reproduced drawings showing his own glass hive. Such objects became something of an exhibit and were shown in public for a small fee as men and women studied bees for moral, technical, and philosophical enlightenment.

It was another kind of observation hive, one that opened like the pages of a book, known as a leaf hive, that was to yield the greatest secrets of the bee to date; the hive had twelve frames, 12 inches high, each containing a single comb. The frames were joined together at the back, hinged as if on the spine of a book so that they could be opened out and examined on both sides. A small piece of comb was put in the top of each frame to help the bees get started. The hive was then closed for three days to encourage them to build further. After this, the frames could be opened like pages and "read" by the observer: it proved to be the best method yet of looking closely at bees.

Most remarkable of all, the leaf hive's Swiss inventor and reader, François Huber, was blind. Huber began to lose his sight when he was fifteen. Despite this, he still managed to marry the girl he loved, who stuck with him against her father's wishes, and to continue his passion for nature. Huber was helped in his work by his servant, François Burnens, who read natural history texts, including those of Swammerdam, to his master, and in the process became absorbed in the subject himself. "This is not the first example of a man who, without education, without wealth, and in the most unfavourable circumstances, was called by nature alone to become a naturalist," wrote Huber.

Using the leaf observation hive, Burnens began to follow Huber's instructions to conduct simple experiments on honeybees. At first the two men carried out experiments that had been performed by others, such as Réaumur. They sought first to validate his findings; and then, by repeating the tests several times, to check their own methods and results. As this work progressed, a covert trial was being carried out by master on servant. Burnens conducted the simple experiments with skill and intelligence. He then went on to more complex ones, and his passion for science grew. At

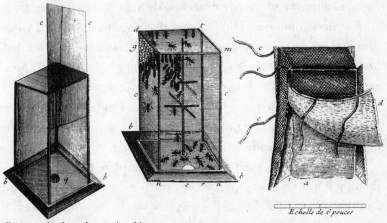

Réaumur's glass observation hives.

the same time, Huber's belief in his servant's powers became stronger and stronger. It got to the point that the blind naturalist had total trust in Burnens: "I hesitated no longer to give him my entire confidence, feeling sure to see well when seeing through his eyes," he wrote.

By this stage, the relationship had evolved from master and servant to that of colleagues. Testing their theories with repeated experiments—the basis of the scientific method—the two men advanced together. Huber described their work in his book *New Observations on Bees*, printed in 1792. These two volumes are easy to understand today, even for the layperson, because the prose is the sum of two people talking to each other.

Huber and Burnens made many discoveries, but the mystery that absorbed them most was the mating of the queen. Swammerdam had proved through his anatomical dissections that the queen was both female and the mother of the hive. But how was she fertilized? The Dutch scientist believed the drone gave out a strong smell and that this was connected to how the queen was fertilized, through some sort of transference. An English naturalist, de Braw, had previously argued that male bees fertilized eggs externally, like fishes or frogs. Réaumur had already dismissed the first theory by reasoning. As for the second idea, Huber knew that eggs were hatched when drones were no longer in the colony, over the winter. They needed, however, to prove that fertilized eggs were laid without drones present in the hive. Burnens got rid of all the drones, and for four days checked through a glass tube at the hive's entrance that none returned. When eggs continued to hatch, they knew that the eggs the queen laid must already be fertilized within her.

Where and how, then, was the queen fertilized? The two men thought it must be connected to the time the drones left the hive for a midday flight. The key to the mystery was to see what happened when the queen also went out at this time.

On June 29, 1788, Huber and Burnens stationed themselves before the hive at eleven o'clock, when the sun had warmed the air. They watched some drones fly out, and then the young queen come to the entrance. "We saw her promenading on the stand of the hive for a few instants, brushing her belly with her posterior legs: neither the bees nor the males that emerged from the hive appeared to bestow any attention upon her," wrote Huber, the fresh exactness of his reportage relaying how Burnens must have described the scene to him.

The queen took flight, moving in horizontal circles 12 or 15 feet above the hive. Then she disappeared for seven minutes. Upon her return, Burnens picked up the queen, examined her abdomen, and found no signs that she had mated. After fifteen minutes back in the hive, the queen emerged again, took off, and flew out of sight. Twenty-seven minutes later, she returned. "We found her then in a very different state from that in which she was after her first excursion," Huber recorded. "The posterior part of her body was filled with a whitish substance, thick and hard, the interior edges of her vulva were covered with it; the vulva itself was partly open and we could readily see that its interior was filled with the same substance."

This, Huber and Burnens reasoned, was the "fecundating liquid" they had seen in the seminal vesicles of the drones. When they opened the hive two days later, the queen's belly was enlarged and she had laid nearly a hundred eggs in the worker cells of the comb. They repeated their vigil several times, with the same results. They had proved that the queen had been fertilized on what we now call the virgin flight.

In the introduction to the first of two volumes of *New Observations on Bees*, Huber pays homage to his faithful servant. The extent of Burnens's work shows a devotion to the subject, as well as to his master; he was clearly compelled by the pursuit of

*François Huber, blind
explorer of the honeybee.*

knowledge. Nowhere is this more evident than in the work he did
with Huber observing the egg-laying workers.

It had already been discovered that worker bees as well as the
queen could lay eggs. But were they just small queens mistaken for
workers, or were there really laying workers? And if so, what sort
of bees did they produce? On August 5, 1788, Burnens and Huber
found eggs and larvae of drones in two hives that had been without
queens for some time. Standing intently in front of the hive,
Burnens tried to spot bees laying, to see if it was workers or a small
queen doing so. Burnens spent eleven days, with scarcely a break,
taking every bee out of the hive individually and checking to see if
it were a queen or a worker. He performed the task with what
Huber calls inconceivable dexterity, taking "the strokes of the
stings" as they came. When he actually saw a worker laying,
Burnens seized her and discovered she had ovaries. They also
found that such laying workers produced only drones. This, we
now know, is because the eggs were unfertilized; an unfertilized
queen will also lay only drone eggs, and the colony will die out.

For fifteen years, Burnens and Huber worked together on bees.
The year after the first volume of *New Observations on Bees* was
printed, they had no fewer than sixty-three hives to work on. By the
time the second volume of the book was published, nineteen years

later, Burnens had left. He had gone, as Huber put it, "to his own people," his rise in life marked by the honor of becoming a magistrate.

Huber's wife, Marie, and son, Pierre, an authority on ants, helped take the place of the servant. It was Pierre who encouraged his father to continue to publish. The second part of *New Observations* explains how wax was not produced from the pollen, as many believed, but was made by the bees themselves.

All Huber's assistants helped him, as he put it, to "pierce the double veil which shrouds, for me, the natural sciences," but it is hard not to notice that he was never so fulsome in his praise of his later helpers as he had been of Burnens.

SCIENTISTS, ARISTOCRATS, FARMERS, and philosophers of the Enlightenment: all admired the insects for their rational, productive ways. Bees and their colonies were adopted as a symbol for the perfect society. The gender of the ruler bee, for example, seemed to change according to who was on the throne. In the court of Charles the Second, the royal beekeeper Moses Rusden argued that the ruling bee in the hive was a king. The frontispiece of his book *A Further Discovery of Bees* (1679) shows a crowned bee, and he states that the "king bee" has the deadliest and most dexterous sting of the hive. Yet Charles Butler had already made it clear that he thought the bee was a queen by the very title of his book *The Feminine Monarchie.*

Bees were recruited as a satirical symbol of a society in an anonymous sixpenny pamphlet, first published in England in 1705, called *The Grumbling Hive: or, Knaves Turn'd Honest.* This was the start of what is now known as *The Fable of the Bees*, one of the most notorious poems of the eighteenth century.

The poem was written by a Dutch physician, Bernard de Mandeville, who had moved to London and worked as a specialist in nervous diseases. The work is a satire based loosely on the concept of the animal fable. De Mandeville had cut his literary teeth by translating fables, including those of La Fontaine. Fables, like allegories, allowed the writer to address human subjects at an oblique angle. In *The Fable of the Bees*, de Mandeville pictures human society as a hive. Instead of being occupied by the most sociable and cooperative of creatures, the bee, the hive is filled with its opposite, the antisocial human.

The gist of the poem's argument is revealed by its subtitle, "Private Vices, Publick Benefits." Vanity, luxury, pride, envy, and prodigality may be examples of man's failings, but they also, de Mandeville mischievously points out, create wealth and provide jobs. Crime keeps multitudes at work: lawyers, jailers, turnkeys, sergeants, bailiffs, locksmiths. There is a disastrous reversal of fortune when the bees turn honest: the hive is ruined. Those who had made money through the corrupt habits of society had now lost their living. "[M]ost writers are always teaching men what they should be, and hardly ever trouble their heads with telling them what they really are," writes de Mandeville.

The poem was republished in 1714 in a longer edition. In 1723, it was expanded again, and the following year published in its final form amid considerable controversy, with a new succession of elucidations and rebuttals included in the text. The growth of the work, from pamphlet poem to a book of close argument and counterargument, shows the ferment of debate that words produced at this time. Such books and pamphlets were relatively new forms at a time when censorship of the written word had been loosened and the minds of the age were pushing at the boundaries of debate.

The poem reads, as does the work of de Mandeville's contemporary Jonathan Swift, like a refreshing squirt of lemon to the intellect, challenging the reader with the sharp sting of its argument. Dr. Samuel Johnson, an English author, said it "opened my views into real life very much." The religious leader John Wesley wrote in his diary: "Til now I imagined there had never appeared in the world such a book as the works of Macheavel. But de Mandeville goes far beyond it." He was consequently seen by some as nothing less than an anti-Christ, and the controversy surrounding *The Fable* continued for the rest of the century. Five years after its publication, ten books had come out attacking it. There were sermons preached against the poem, and letters to the press denouncing its content. The Grand Jury of Middlesex put the book forward as a public nuisance. The controversy spread as it was translated into French and German. In France the book was ordered to be burned by the common hangman. Wherever the fable was read, the hive of society buzzed as if attacked by an intruder.

BEES AND HIVES were also brought to an aristocratic audience through showmanship. Thomas Wildman (1734–1781) has been called the Barnum of Beekeepers for his show-stopping displays, which he took to the English court and beyond. In his insect spectaculars, Wildman was carried through London on a chair covered in bees; he enacted a battle between bees and three mastiffs; he got the bees to fly from one place to another, as if he were a conductor leading tens of thousands of flying half notes; and he trotted on a horse, followed by swarms that settled on him as he rode.

These exuberant exhibitions were founded on a more serious knowledge of how bees worked. Wildman's methods for beekeeping were laid out in his book *A Treatise on the Management of Bees* (1768). The book begins with a list of five hundred subscribers, showing how Wildman popularized the honeybee from King George III and his wife, Queen Charlotte, to whom the book is dedicated, to members of the Royal Society, dukes, and tradesmen. At the heart of his work was the same quest that still engaged the mind of many beekeepers: how to manage the bees without killing them. "Were we to kill the hen for her egg, the cow for her milk, or the sheep for the fleece it bears," he writes, "everyone would instantly see how much we should act contrary to our own interests: and yet this is practised every year, in our inhuman and impolitic slaughter of the bees."

Thomas Wildman was born in Devon, one of England's great beekeeping counties. Although he made and sold wooden hives, his main method of beekeeping was to use four or more flat-topped skeps piled on top of each other for one colony. He could take off the top skep and extract the honey, rather as a super is used in a modern system, rotating and removing the skeps in the tower as needed. He used some of the honey to make mead, which he preferred dry, fermenting the sweetness right out to get "a fine, racy flavour."

Wildman sold his bee expertise not just through his shows and book but by going around the country visiting people to advise them on their apiaries. He thought that the bees should be "near the mansion-house, on account of the convenience of watching them."

Wildman's nephew Daniel was operating at the same time in London, and was a professional bee-equipment maker with a shop at 326 Holborn. He kept bees on top of his house and, in order to see how far his bees were flying to forage, he marked some with flour and discovered that they were going right up to Hampstead Heath. Daniel Wildman, a good businessman like his uncle, also wrote a beekeeping manual, entitled *A compleat guide for the management of bees*, which came out in 1773 and went into many editions, including a French translation. He traveled around the continent giving exhibitions, and there are records of his nightly exhibitions of bees at the Jubilee Gardens, Islington, from June 20, 1772. An interest in bees had clearly spread from the court and countryside to the general urban public.

~

AS WELL AS BOOKS ON BEES, the eighteenth century has left us an early piece of writing specifically on honey by a Covent Garden apothecary, Sir John Hill. This delightful tract from 1759 begins with a theme that holds true today: "The slight regard at this time paid to the medicinal virtues of Honey," Hill writes, "is an instance of the neglect men shew to common objects, whatever be their value." In other words, we look down on that which is under our nose. Honey is a useful treatment for many ailments, he argues. It can, for example, help loosen tough phlegm, a common nuisance and easily cured. Keep honey by your bedside, Hill advises, and take a spoonful last thing at night "letting it go gently down." Use it again the next day, too, and continue this regime until the symptoms improve. He believed honey relieved hoarseness, coughs, asthma, and to some degree (less plausibly) consumption, if caught early enough.

Hill is specific about the best honeys to use for medicinal problems: for English honey, the springtime harvest is best, because the bees are most vigorous and capture the full force of the first flowers. He also writes of imported French, Italian, and Swiss honeys, adding that honeys similar to those of Hymettus and Hybla could be found in England where the bees are foraging in the same plants. For example, he cites a dell on the left-hand side of the road heading from Denham to Rickmansworth, which is fragrant in the evening air because of its wild thyme. The honey from this source, he says, is "perfectly Hyblaean" in its delicate sweetness and quality.

~

WITH ALL THE political turbulence of the seventeenth and eighteenth centuries, it is remarkable how the honeybee continued to be seen as such a positive symbol throughout the period. Its ruling queen and obedient servants survived regicide and revolution. Seen as an emblem of monarchy, the honeybee became the sign of an emperor: Napoleon Bonaparte took the Bourbons' fleur-de-lis, turned it upside down, and transposed it into a bee. His coronation robe was covered with the insect. Marie Tussaud, the famous modeler, taught Louis XVI's sister Elisabeth to mold beeswax; later Tussaud made the death masks of the guillotined king and queen in the same material. People had to adapt to survive, as did the bees—at least as far as the way they were perceived. As the nineteenth century began, the symbolism and use of the honeybee was set to change once more, to move onward with science into the industrial age.

CHAPTER SEVEN

FRONTIERS

The honeybee is not indigenous to North America. When the seventeenth-century Puritan missionary John Eliot translated the Bible into the Algonquian language, he found no words for *honeybee* and *honey*, and although Columbus mentions wax, this would have come from the stingless bees native to the Americas. The European settlers brought with them the dark honeybee of northern Europe, and this was the first race to inhabit the East Coast. Colonies were shipped across the Atlantic, alongside the cattle and other livestock, to settle, reproduce, and feed the people now digging into the edge of this new land.

The first settler bees were probably not on the mainland but offshore. In 1609, *The Sea Adventurer*, a ship of the Virginia Company of London, was wrecked off the most northerly island of the Bermudas en route to the recent settlement of Jamestown in Virginia. (William Shakespeare was later to use this wreck as part of the inspiration for *The Tempest*.) Subsequent colonizers of the Bermudas brought plants, goats, cattle, and honeybees. On May 25, 1617, the Earl of Warwick, patron of the Virginia Company, reported to his brother, Sir Nathaniel Rich: "The bees that you sent doe prosper very well."

The first record of honeybees on the North American mainland dates back to 1622. The previous year, four masters of the Virginia

Company had been commissioned to convey settlers and goods to Virginia from England. In return for this service, they were allowed to fish off the coast, and they also traded for fur. By May 1622, three of the ships had among them delivered ninety settlers. At least two of these ships brought beehives, alongside seed, fruit trees, pigeons, and mastiffs.

The very first Europeans arriving in the New World hoped to feed on native plants and soon learned about some of the native American crops, particularly corn, beans, and squash. Those Pilgrim Fathers who survived the grim and death-filled winter of their arrival celebrated a bumper crop of maize, and other bounty, with the first Thanksgiving feast, eating with the Wampanoag Native Americans who had helped them. But settlers soon discovered that it was safer to provide for themselves, and agriculture was transplanted from the Old World to the New. They grew wheat, barley, and oats and kept honeybees alongside cattle, pigs, and sheep. The Native Americans knew how to extract the rising sap of the maple and boil it down to a sweet syrup, but the early settlers imported their familiar Old World sweetness by bringing over honeybees.

A number of colonies of bees came along with the twenty thousand Englishmen and -women who undertook the arduous voyage to an improved, new England in the 1630s. The town of Newbury, Massachusetts, founded in 1635, had a communal apiary five years later, run by a beekeeper named Eales. A Native American watching the bees working, having previously seen the arrival of the horse and the ox, wondered at the way the settlers put their animals to work. According to Frank Pellet in his 1938 history of American beekeeping, he commented: "Huh! White man work, make horse work, make ox work, make fly work: this Injun go away." All the same, beekeeping does not appear to have been a profitable trade for Eales. The town subsidized his hive-making but he still went on to become the first town pauper. A "stok" of bees (the words *stad*, *stok*, *stake*, *stall*, or *skep* were all used for a colony) was worth the equivalent of fifteen days' manual labor in the 1640s, not including the trouble required to maintain it, so perhaps the price of sweetness was too high.

The insect fared better elsewhere. Gathering research from around the world for his book *The Reformed Commonwealth of Bees*, Samuel Hartlib in 1655 noted that "bees thrive very much in New England." In the Swedish settlements in Pennsylvania, it was reported that "bees thrive and multiply exceedingly . . . the Swedes often get great store of them in the wood where they are free from anybody." Bees brought to Boston in 1670 were said to have "spread over the continent." This was an exaggeration: it is true that the honeybee eventually covered America—but from various sources, and not quite yet.

Yet wild honey was recorded as plentiful in the Carolinas in the early eighteenth century, and bees were common in the cypress swamps of Florida in 1765, where great quantities of honey and wax were used by both Native Americans and settlers. By the time of the War of Independence ten years later, a British Army officer passing through Pennsylvania commented that "almost every farmhouse has 7 or 8 hives of bees." One of the first banknotes issued by Congress at this time depicted two straw skeps in a shelter: it was a symbol of the hard work, thrift, and enterprise that were needed to make this young land great.

So the honeybee did spread, both by the natural means of swarming and with human help. When the bees swarmed, they found new patches to forage, flying at least a little ahead of the settlers, and advancing into the territory occupied by native peoples. In his *Notes on the State of Virginia* (1784), Thomas Jefferson wrote: "The Indians . . . call them the white man's fly, and consider their approach as indicating the approach of the settlements of the whites." It was said that as the bee advanced, the Indian and the buffalo retired.

Dr. Everett Oertel, who mapped the migration of the honeybee across America, calculated that their nests moved along the Missouri River at a rate of 600 miles per fourteen years in the 1800s. They thrived best on land bordering both woods and prairies; when the trees stopped flowering in the spring and summer, the insects could fly to the plants on more open ground.

Native Americans, highly skilled at tracking and hunting wild food, soon became adept at harvesting this new source of sweetness. Settlers arriving in Wisconsin in the 1820s found bee trees with ladders left next to them so the natives could plunder the nests. Some tribes renewed the land by burning it periodically, encouraging the growth of grasses so that deer would come to feed in the area. This practice left burnt-out hollow trees, which could easily be settled by swarms. One legacy of the abundance of honey consequently found in Wisconsin was an unusually large number of bee- and honey-related names, including a Honey Island, three Honey Lakes, four Honey Creeks, Bee Bluff, Bee Hollow, and a number of Beetowns.

Once a nest was found, the honey hunters pacified the bees with smoke, then cut out the section of tree with a nest, either to set it up as a simple log hive closer to home, or to extract the honeycomb on the spot. In 1847, one such Wisconsin trip was described: "Parties go bee hunting for months together in Summer, they take wagons and a pair of Oxen, an ax and coffeepot, and that's all except barrels for the honey. When they come to a prairie they turn out the cattle, and if they locate a bee tree, they chop it down, smother the bees and take the honey, barrel it up, then *ditto* several times a day perhaps. They shoot for meat, roast corn in a frying pan for coffee, barter honey for flour from settlers, bake it in a pan, and sleep in their wagons at night."

The methods used to track down the bees in the woods were various and often ingenious. Paul Dudley, in Massachusetts in 1721, instructed hunters "to set out on a clear sun-shiny day, with pocket compass, rule and a sheet of paper." His book, an account of a method found in New England for discovering where the bees hive in the woods in order to get their honey, advises putting honey on a plate or trencher and releasing a bee that has been caught foraging. Mark the line it takes on a piece of paper. Then go to another spot, not far away, and release another bee. The two angles marked together will help the hunter find the direction of the tree.

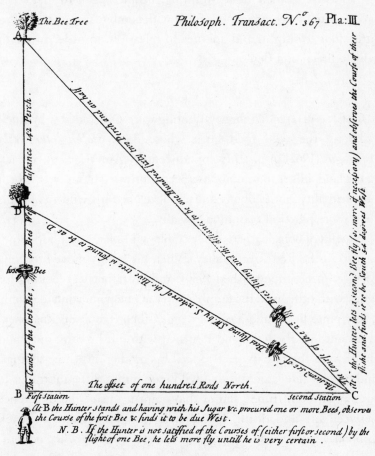

How to track down wild honeybees in eighteenth-century America.

On the other hand, Patrick Campbell, in his *Travels in the Interior Inhabited Parts of North America in the years 1791 and 1792*, recommends putting a flat stone with some wax on it over a fire. Close by the wax, put some more honey and vermilion pigment. The bee is drawn to the stone by the smell of beeswax, goes to the honey, and is marked by the red. If you watch in which direction it flies off and time its return, it is possible to work out both the direction and distance of the honey tree.

To assist in their honey-hunting, people began to make bee boxes, with one compartment filled with comb to attract the bee, a little door to trap it, and another to release it, in order to see in which direction it flew.

~

SUCH METHODS OF honey-hunting were later added to by Euell Gibbons, the wild food guru whose book *Stalking the Wild Asparagus* (1962) inspired twentieth-century Americans to go back to the land. His readers may have been motivated by an earthy form of spirituality, but Gibbons himself developed his wild-food skills for a more practical reason: to stay alive.

Euell Gibbons learned about edible wild plants as a boy growing up in the Red River Valley. When his family moved to New Mexico in the impoverished 1930s, and his father was looking for work with little luck, the family pantry at one point contained only a few pinto beans and a solitary egg. Gibbons took his knapsack, went out into the wild, and returned with mushrooms, nuts, and the fruit of the prickly pear to feed his mother, his three siblings, and himself. He later said that wild food meant different things to him at different times, but at that point it meant the difference between life and death. While struggling to become a writer, Gibbons lived on wild food for five years; it was only when he combined his subsistence skills with his writing that he had a hit.

Stalking the Wild Asparagus recorded his foraging experiences, from swamplands to downtown, including the fifteen species of edible plants he found on a vacant lot in Chicago. The book was a bestseller and has become a classic.

Gibbons starts his chapter on wild honey by describing his grandfather's exceptional skills with feral bees. Most bee raiders dress up in veils and other protective clothing; Gibbons's grandfather would merely open his shirt collar, roll up his sleeves, and get to work. Sometimes, he went to the woods just to collect wild honey; at others, he brought the bees back as well, and installed them in a hive closer to home.

When he found a colony of wild bees in a hollow tree, the old man would first drill a hole and stuff it with burning rags, before splitting off a section of the wood to expose the nest. He would take out the comb, holding it up to the light to distinguish between the light honey cells and those that were darker with the brood and pollen stores. Gibbons recalls on one occasion watching his grandfather emerge from a fog of smoke and bees, moving in his quiet, deliberate way to bring him a pale piece of honeycomb that the bees had made from cotton blossom; the bees were crawling so densely over the old man's glasses that he had to light his pipe to keep them off his face.

So it was only a matter of time before Euell Gibbons tracked down wild honey for himself. He found a great deal just by discovering nests by accident, or hearing about bees that had colonized buildings. Later, he decided to test a more "scientific" method of tracking them down while on a camping trip. Although he had read about the elaborate bee boxes that were used to trap bees, in the end he improvised with a cobbled-together kit using an aluminum cake cover, an old piece of honeycomb, and some blue carpenter's chalk.

First, Gibbons filled the old, honeyless comb with a sugar solution that had been scented with a little anise oil. He then put the comb on a stone—and went back to his fishing. Half an hour later, he returned to find the bees had deserted their flowers for this attractively scented food. He saw the insects fly back toward their colony, watching for the flash of their wings in the sunlight for as long as possible, and noting the direction they took. This was the beeline that would, eventually, lead to the nest.

The next task was to gauge the distance to the nest. With great care, Gibbons daubed a foraging bee with a little of his blue carpenter's chalk, which he had dissolved in a drop of water and put on the end of a camel-hair brush. He waited with baited breath until the same blue-bottomed bee returned six minutes later. According to an article he had read by G. H. Edgell, "Bee Hunter," published in *Atlantic Monthly* in 1949, this meant the colony was less than a mile away.

Although bees take as direct a route as possible between their nest and foraging sites, there will be obstacles to avoid and diversions to make. It was necessary, therefore, for Gibbons to follow them in short stages. To do this, he carefully trapped the bees on the comb with the aluminum cake cover. Walking along the beeline, Gibbons advanced with the insects in the improvised box. Then he uncovered the comb to let the bees fly off. But they zigzagged and double-eighted and flew off helter-skelter: the line was lost. Gibbons refilled the comb, put on some more anise solution, and waited for ten anxious minutes. Fortunately, the bees came back, and their line toward the nest was then tracked in three more hops. When Gibbons noticed the insects were going back from whence he'd come, he knew the bee tree had been passed. Looking up for a likely site, he spotted a beech tree; the mere flash of a bee wing on a bough, reflecting the light of the setting sun, revealed the site of the stash. The search was over. The next day, Gibbons took two pailfuls of honey out of the tree, in the process receiving just a single sting—in his bottom as he bent over to cut out the comb.

This trip turned out to be beginner's luck, however. Euell Gibbons used his homemade bee kit again, with some success; but he would often lose the trail or the bees. Patience, he summarized, is what the bee hunter most needs—and, clearly, blue carpenter's chalk, old honeycomb, and an aluminum cake cover can come in handy.

BY THE EARLY nineteenth century, the honeybee in America had spread as far as it could from its original landing points. When the insects encountered the Appalachians and Allegheny mountains, their easy progress was thwarted. But they continued by other means, carried along with human pioneers: as the new Americans went westward to find freedom and fortune, the honeybee went too.

Some of the insects flew on with the followers of the Church of Jesus Christ of Latter-day Saints, commonly known as Mormons. Their faith's founder, Joseph Smith, said that as a farm boy in upstate New York he was directed to inscribed golden plates by the angel Moroni, and in a series of revelations translated their words. The text was published in 1830 as *The Book of Mormon*, which describes ancient Israelite tribes who came over to America, centuries before Christ, and underwent experiences similar to those found in the Old Testament. In language reminiscent of the King James Bible, the book describes how bees were part of the bounty brought to this new land: "And they did also carry with them deseret, which, by interpretation, is a honey bee; and thus they did carry with them swarms of bees, and all manner of that which was upon the face of the land, seeds of every kind."

Deseret—as the honeybee was referred to in *The Book of Mormon*—was important to Joseph Smith in other ways. He had a keen eye for how symbolism could bring and bind people together.

The bees' ordered society was an example of cooperation and productivity; Mormon businesses were all cooperatives—unusual in the free-for-all of the frontier. The reproductive life of the hive, with its ruler mated by several drones, perhaps also reinforced the Mormons' practice of polygamy—a custom that was a major reason for the sect's unpopularity in its early days. Joseph Smith was eventually lynched by a mob, such was the animosity toward the group. It was said his grave was protected by hives of bees, put there to prevent grave robbers from digging up and desecrating his body.

In order to build their society in a place free from persecution, some 66,000 Mormons set out on their epic journey toward the Salt Lake Basin, following their new leader, Brigham Young. The Mormons arrived in 1847 in what was to become Utah and Salt Lake City. Honeybees were first transported there in the back of a covered wagon the following year, and by 1851, several hives were said to be responding well to local conditions. The land was won from the Mexicans in 1848, and at this point the Mormons called the territory Deseret, after the honeybee. The 1850 territorial seal is based on a beehive, as a symbol of organization, unity, and productivity. However, when the territory became the forty-fifth state in 1896, after the Mormons had renounced polygamy, the name Utah was picked, instead, after the Ute Native Americans of the area.

The Mormons' honeybee symbolism continued in their choice of house for the leader. The building has a roof based on the shape of a skep, and was called the Beehive. Brigham Young, the first governor of the territory of Deseret, lived here from 1854 until his death in 1877. It was restored in 1961 and can now be visited by the public. Deseret Telegraph Company connected all the Mormon settlements to the Beehive house; communication was again reinforced by the symbolism of the honeybee, which is able to tell its fellow insects about nectar sources and other business of the hive.

〜

TRANSPORTING HONEYBEES around the world was no easy matter. The early journeys from England to America took between one and two months, and as late as the nineteenth century, travels to New Zealand and Australia could take five or six months. To keep the bees alive during such long trips, it was important to keep them cool. If the bees were in their winter cluster, they had a greater chance of surviving. This slowed down their metabolic rate, in which state they could make the journey in a quiet, stable condition.

One of the pioneer beekeepers of New Zealand was William Charles Cotton, a Victorian whose adventures around the globe read like a storybook. From the start, Cotton's imagination was unorthodox. As a boy, not content merely to hear about the supposedly "ox-born" bees in Virgil's *Georgics, Book Four,* he set out to repeat the experiment himself. As you may recall, this involves bludgeoning an ox to death in a closed room. "I suppose I was born an experimentalyst," Cotton later wrote in *My Bee Book* (1842), "so I went out next morning with a full determination to try a grand one. I found a shed which would do nicely, which had all that Virgil requires. I had no pity for the poor cow—no, not I—when a swarm of Bees was to be the glorious result: she would surely, I thought, be happy in her death, as she would give life to so many glorious creatures." Fortunately, the experiment went no further—Cotton's father got wind of his son's attempt to obtain a cow by bribing a farmer with the promise of honey, and bought the boy his first stock of bees instead.

The anecdote indicates something of Cotton's adult character, albeit with the bloodthirstiness of boyhood replaced by kinder traits. His early life followed an apparently orthodox route—Eton, Christ Church, a career in the church—but, as with the Virgil episode, you often sense a certain excitability buzzing in the background. His temperament greatly concerned his father, a governor of the Bank of England and a more sober citizen. Cotton senior certainly disapproved of his son accompanying his mentor, Dr. George

August Selwyn, the newly appointed bishop of New Zealand, to become chaplain and teacher of a collegiate-style school, St. John's, near the Bay of Islands on the North Island.

Despite his father's unease, the young cleric was on board the *Tomatin* on Boxing Day 1841, when she set sail from Plymouth. Alongside "a goodly fellowship of emigrants, schoolmasters, deacons, and priests, *with a Bishop at their head*," the cargo contained many thousands of bees. How did Cotton intend to pack his bees for this journey to the other side of the world? Writing before his departure, he proudly described his plans. His insects were to be stored by four methods. Some skeps would be in hogsheads—recycled wine barrels—packed with ice. So far, so good; then the elaborations of the arrangements start to read more like the plans of an inventive mind rather than the simple effectiveness of a practical one. Once they crossed the equator, Cotton was to let the melted water out and measure it to calculate how much ice would be left to keep the bees cool. The hogsheads had breathing holes for ventilation, with the tubes covered in perforated zinc to stop the bees from escaping. He planned to keep another hive cool by means of evaporation, surrounding the skep with running water. This hive would be mounted on springs so

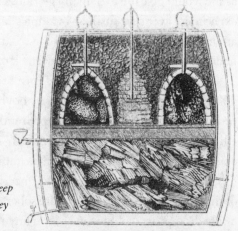

William Cotton's barrel of bees, with the skeps kept over ice to keep the insects sleepy and calm as they voyaged to New Zealand.

the motion of the ship would not disturb the bees. For the same reason, another hive was to be kept on a set of gimbles, the device that keeps a ship's compass on the level despite the motion of the sea. Furthermore, Cotton wanted an observation hive in his cabin. These bees would be active, and he would feed them with honey.

It is uncertain how, in fact, the bees were transported and, indeed, whether they arrived dead or alive. The Devon beekeeper Thomas Woodbury, writing in 1858, says Cotton's hives had been thrown overboard—"to the indescribable grief and disappointment of their amiable and enthusiastic owner"—by superstitious sailors who thought the insects were to blame for storms and bad luck on the latter part of the voyage. Cotton's diaries do not describe this event, which is surprising since he recorded other parts of his trip in gleeful detail. (He tells, for example, of the sharks who came to bite at the salt pork the sailors were towing over the bows so as to wash off some of its brine: he felt a frisson of danger, since he had recently taken a refreshing dip, assured that sharks did not swim so far south.) Did the sailors toss the hives into the sea, or not? Peter Barrett, the Australian beekeeper who has written two meticulously researched volumes about the introduction of the honeybee to Australasia, *The Immigrant Bees*, as well as a volume specifically on Cotton, has unearthed reported evidence supporting both outcomes.

What *is* certain is that once he arrived in New Zealand, Cotton was supplied with some bees from Australia. Honeybees were first taken to New South Wales in 1822. Previously, the Aborigines had collected honey, but this was from stingless bees. They would track down a nest by fastening a speck of white down to the back of an insect with a dot of gum, and follow it to the colony. They called the new bee "the white-fellow's sugar bag." (Another, tangential, historical detail illustrates the times: in 1829, convict number 680 was a woman from Gloucester—transported for the offense of stealing bees' honey.)

There were no native bees of any kind in New Zealand, but the honeybee had already been imported by the time Cotton landed. *Apis mellifera* was first brought to the North Island in 1839 by Miss Mary Anna Bumby, from Thirsk in North Yorkshire, who came over with her bees to become housekeeper to her missionary brother. She put her two straw skeps of bees in the mission church-yard, out of the way of curious Maori.

As well as leading a physically active life in New Zealand—swimming, riding, sailing, and walking—Cotton kept his beloved bees and supplied honey to the college. He had such a connection with these insects that he was said to be able to walk around with one in his pocket. As regards his theories of beekeeping, he dis-approved entirely of the method of collecting honey by killing the bees, and writes on the subject with characteristic passion: "NEVER KILL YOUR BEES . . . every one of you must feel some sorrow when you *murder* by thousands in the autumn those who have worked hard for you all the summer, and are ready to do so again next year."

Honeybees fly through the rest of Cotton's lively diaries and the letters he wrote to his sisters in England. These missives contain many colorful anecdotes of New World life, put in the envelopes alongside parrot feathers and illustrations of Maori tattoos. Cotton taught the Maori to keep bees, introducing them to honey by dip-ping his finger into a plateful and offering them a taste. "The uni-versal expression of admiration," he records, "is *He mea uka wakaharahara* 'a very exceeding sweet thing,' the last word, the highest superlative, pronounced with great energy."

But although Cotton was the life and soul of the settler com-munity, his animation had a darker side. "Cheery and lively but an anxiety, from time to time, truly," was the heartfelt summation of Sarah Selwyn, the wife of Cotton's bishop. He had, she said, a want of ballast to steady his eccentricities. Despite his goodness and inventive enthusiasms, one was uncertain what he would do next, she said.

Alongside the delight Cotton took in his honeybees, they also embodied his more dejected mental states. His letters reveal the abject pain of unrequited love, when it emerged that the woman he loved, but had left behind in England, would not await his return. His feelings were exacerbated by his sheer physical distance from her. In a letter to Phoebe, his favorite sister and closest confidant, Cotton wrote: "Mrs Dudley kindly played to me some tune on Mrs Selwyn's piano . . . and big silent tears rolled down my face. . . . As an emblem I suppose of my blighted hopes—the swarm of bees with which I have been occupied that day would not stay, but flew off and took up their dwelling in the bush."

On the day of Cotton's departure from New Zealand, six years after his arrival, three swarms of bees came out "as if to bid farewell." He returned at his father's behest. Cotton's spirit did not fit in with Victorian England. If his trip to the New World sounds like a liberation, his return was the opposite. He brought his way of thinking and behaving back to a more formal world where he could no longer ride his horse General, or sail across the Bay of Islands, or teach the Maori people about honey and beekeeping. Instead, he had the more somber realities of a wifeless life administering a parish in Cheshire. His love of bees continued, and he attended the first meeting of the British Beekeepers' Association. But there were troubles, too. As well as suffering much mental turmoil, his finances ran amok—much to the anxiety of his banker father—and then there was the death of his dearest sister, Phoebe. Unable to function, Cotton died in Chiswick in 1879, aged sixty-six, in a humane Quaker-run asylum; he was buried in St. John's, Leytonstone.

William Cotton remains, for all his personal woes, one of the most attractive characters in the history of the beekeeping world, carrying the breeze of a New World freedom through the pages of his journal. Among his books, *A Manual for New Zealand Beekeepers* (1848) and, in Maori, *Ko Nga Pi* (*Treatise on Bees*), printed by the St. John's College Press in New Zealand in 1849, were pioneering works.

Cotton's obituary in the *British Bee Journal* in January 1880 underlines one legacy of early beekeeping in New Zealand that remains important to the country's economy and honey production even today: "Before the introduction of the honey-bee into New Zealand, they had to send over to England every year for the white clover seed as it did not seed freely there, but by the agency of the bees they are now able to export it. New Zealand is such a good country for bees, that Mr Cotton told me, one stock had increased to twenty-six in one year." This last statistic sounds like a final note of characteristically excitable exaggeration.

But the honeybee *did* do well in the New World; as the honey expert Dr. Eva Crane points out, it can do better there than in the countries where it had evolved—California and New Zealand are famous for their honeys.

CHAPTER EIGHT

FOLKLORE AND SCIENCE

Let's be fanciful for a moment and compare a colony of bees to the questing human mind. Our wandering thoughts fly off in a thousand directions; then these winged notions return to the hive of the head to make honey. The sweet and strange stories of bee folklore contain nectar from some of the wilder flowers of the field; yet the large quantity of such stories and beliefs also reflects our longstanding preoccupation with these mysterious insects. These tales often contain an element of sound advice, too.

The same superstitions cropped up time after time in the eighteenth and nineteenth centuries. All over Europe, it was deemed bad luck to buy a colony; rather, they should be acquired by bartering. In 1720, Thomas Lupton's *Thousand Notable Things* includes the comment: "If you have no Stocks of Bees, but must buy them, I advise you first, not to give Money for them, but some other Commodity; for though there can be nothing in it but a superstitious Observation, yet things often dishearten People that are apt to credit such Reports." Country folk exchanged goods such as wheat, barley, and oats for their bees, and in mid-nineteenth-century Hampshire, a colony was worth a small pig. Beliefs about payment for bees continued, in various forms, into the twentieth century. In the first decade of the 1900s, in Sussex, it was reputedly acceptable to use money—but only gold; in Devon as late as the 1930s, beekeepers were warned against being given bees rather than buying or bartering for them; and in 1948, in Surrey, one woman urged: "If you buy bees, you *must* give silver for them." Perhaps all this care about payment derives from peasant thriftiness; in France, in the Vosges, it was considered bad luck—and presumably, bad judgment—to pay too *much* for bees.

Beliefs about bees often reflect the way they were seen as moral creatures; any disruption in the household would provoke a reaction in the colony. Bad language and quarreling would offend the bees: such behavior could result in a stinging punishment from the reproving insects. Bedfordshire beekeepers would sing psalms before their hives if the bees were not thriving. In France, it was believed that a sting was a message from a relative languishing in Purgatory, a sharp reminder of the wages of sin. On a secular level—though no less fantastically—Central European peasants had a custom of giving their bees written contracts, promising to look after them throughout the year, hoping in turn to be repaid by productivity.

Swarming bees were seen as an omen, presaging some important event. If they landed on a dead branch or a hedge stake, a death could be imminent; if they flew into a house, a stranger would arrive; if they landed on a roof, good luck was on the way (perhaps in the form of a local source of honey, if only some brave person took the chance of collecting it).

Bees were part of the family, so important events such as marriage and death had—of course—to be reported to them, a ritual known as "telling the bees." This was first recorded in England in the seventeenth century and became more common from around 1800. The various customs included tapping the hive with a key, whispering the news to the insects, and leaving an appropriate gift—a piece of wedding cake or funeral biscuits dipped in wine—at the hive's entrance. If the bees were not kept informed of events,

"Telling the bees."

they might fly away. This superstition has a modicum of sense: when a beekeeper died, his skills went with him; how the insects were treated would perhaps indicate their future fortunes.

It was also the custom to turn hives away from the beekeeper's coffin as it was carried out of the house. In their *Dictionary of Superstitions*, Iona Opie and Moira Tatem relate one late-eighteenth-century Devon funeral when chaos ensued because the bees were treated incorrectly: "[A]s the Corpse was placed in the Hearse, and the horsemen . . . were drawn up in order for the procession of the Funeral, a person called out, 'turn the Bees,' when a Servant who had no knowledge of such a Custom, instead of turning the Hives about, lifted them up, and laid them down on their sides. The Bees, thus hastily invaded, instantly attacked and fastened on the Horses and their Riders. It was in vain they galloped off, the Bees as precipitately followed [and] a general Confusion took place, attended with loss of Hats, Wigs, &c."

Alongside the custom of "telling the bees," the hives might be draped in black crepe, or with a piece of black wool, following a death in the family. A newspaper article in 1925 related how one Worcestershire woman would dress up—including gloves—to inform the bees of important family news. After a death, she wore widow's black; for a wedding, "she donned her gayest dress and carried white ribbon"; for a birth, the ribbon would be pink or blue.

In an echo of ancient beliefs, some saw the bees as the embodiment of human souls. *Lincolnshire Notes and Queries* (1851) tells the story of two traveling servants from the start of that century: "[They] laid down by the road-side to rest, and one fell asleep. The other, seeing a bee settle on a neighbouring wall and go into a little hole, put the end of his staff in the hole, and so imprisoned the bee. Wishing to pursue his journey, he endeavoured to awaken his companion but was unable to do so, till, resuming his stick, the bee flew to the sleeping man and went into his ear. His companion then awoke him, remarking how soundly he had been asleep and asked what he had been dreaming of—'Oh!' said he, 'I dreamt that you shut me up in a dark cave, and I could not awake until you let me out.'"

AGAINST THE BACKDROP of such whimsy came the rising rationality of the eighteenth and nineteenth centuries, with scientists advocating proof over blind belief. Pioneering scientists, particularly the Swedish naturalist Carolus Linnaeus (1707–1778), began to classify insects as a separate branch of natural history; with such thinking came the rise of the entomologist. Nonetheless, insects were still widely seen as a curious subject for scientific study. When the Reverend William Kirby, a Suffolk vicar, listed three hundred species in his 1802 monograph on English bees, he clearly faced some derision. While the botanist is admired for studying mosses and lichen, he wrote the physical smallness of these creatures meant that "an Entomologist is synonymous with every thing futile and childish"—in effect, they were boys chasing bugs. In his later *Introduction to Entomology* (1815) Kirby is at pains to champion the honeybee: "Of all the insect associations," he writes, "there are none that have more excited the attention and admiration of mankind in every age, or been more universally interesting, than the colonies of these little, useful creatures."

But by the time of the Victorians, insects were avidly collected. Specialist groups sprung up, most notably the Entomological Society of London, of which Charles Darwin was a lifelong member. During the summer months, working-class men would find rare species and sell them to enthusiasts. The natural history writer David Elliston has suggested this rise in interest was perhaps a symbol of the new urban middle classes' need for nature; trapped in their new towns and cities, these fledgling city dwellers needed a memory of freedom and flight.

Writers on bees tended to divide into those who were absorbed by the science and those who were commercial beekeepers, who were often down-to-earth people making a living in a rural economy. Both sides had much to learn from each other in this age of improvement: beekeepers found applications for the scientific theories; and those exploring the science—frequently clergymen—were beekeepers themselves and therefore practical in bent, if not explicitly commercial. This crossover between the science and practice was a key aspect of the nineteenth century, especially after the 1850s, when production shifted up a gear and beekeeping moved from cottage industry to factory production. The honeybee became business—and in the United States, big business.

At the start of the nineteenth century, the challenge was to promote a more rational form of beekeeping. A Nottingham skeppist, Robert Huish, published his forthright views on bees, gathered from his experience of keeping up to a hundred hives, in the journal *Gardener, Florist and Apiculturist*. He scorned superstitions such as the idea that a colony bought with money would not thrive.

"Excepting the Spanish," he wrote, "I know of no nation which entertains such superstitious prejudices, in regard to bees, as the English."

In hindsight, of course, it is easy to see this debunking author's own mistakes. Huish thought the idea that bees mated in the air quite absurd, and believed that wax was collected from plants, like pollen and nectar, rather than made by the bee. He thought bees tended by women were bad tempered, and blamed the poor state of the bees he saw in one Sussex village on the gender of their keepers. But whatever his misconceptions, Huish was genuinely trying to get rid of some of the more wide-and-wayward ideas.

The push was on to convince more cottagers to keep bees. The English radical William Cobbett tried to educate the common man, woman, and child to live productively at a frugal level, to make, by skill and graft, a decent life and living that could help the family make the slow, steady ascent up the social scale. Cobbett was a propagandist for the practical, and his *Cottage Economy* (1822) adds the muscles of exhortation to the bones of instruction. The honeybee fitted Cobbett's purposes admirably. He saw an educational and moral purpose in maintaining livestock such as bees; but above all it offered the cottager the chance, for no cost other than his own labor, to make something from nothing. "He must be a stupid countryman indeed who cannot make a bee-hive; and a lazy one indeed if he will not, if he can," he wrote. "In short there is nothing but care demanded and there are very few situations in the country, especially in the south of England, where our labouring man may not have half a dozen stalls of bees to take every year."

To kill or not to kill the bees? This question continued to vex beekeepers. We know William Cotton's vehement opposition to the practice; before he departed for New Zealand, he specifically addressed cottagers with this advice. Cobbett, however, thought sparing the bees when the honey was collected was mere whimsy;

individual bees would in any case perish from age, and the less strong colonies would die over winter. Another apiarist author, Richard Smith, also scorned the idea of not killing bees, warning that a patch could be overstocked with bees, just as pasture could with cattle.

Another controversy hinged on the skep versus the wooden hive. Many early-nineteenth-century beekeepers preferred straw hives to wooden boxes. They thought straw provided better insulation and more protection from drought and rain; they were also cheaper than wooden hives, so better suited to the cottager. The best hives, said Cobbett, were made from rye straw, topped with thatch to keep out the rain. Each swarm at the outset should be housed in a new hive, because used ones could harbor moths and other problems and diseases.

One British author with advanced ideas on this subject was Dr. Edward Bevan. His delightful 1827 book, *The Honeybee, its Natural History, Physiology, and Management* (the second edition was dedicated to the current "queen bee," Victoria), successfully combined the practical with the historical and scientific. He kept only half a dozen hives, mostly for observation, but he saw how bees could be both a profitable part of cottage economy and a source of "pleasing and rational" amusement for the man of leisure. The bee, he said, "tends to enlarge and harmonize the mind, and to elevate it into worthy conceptions of Nature and its Author." His own writing transports you with the bees toward plants; he writes evocatively, for example, of the loud humming in the ivy-mantled tower of an old castle.

In the debate about straw skeps versus wooden hives, Dr. Bevan came down on the side of wood. The system of building up layers of boxes, or "storyfying," so that the honey could be removed without taking away all the bees, worked better in wooden hives. "I think wooden boxes have a great superiority over straw hives; they are more firm and steady, better suited for observing the operations of the bees through the glass windows in the backs and sides, and less liable to harbour moths, spiders, and other insects," he wrote.

Within a few years, the manufactured wooden hive began to replace the homemade straw skep, and this led to large increases in honey yields. The drive to produce more honey, through the application of science, was to be the story of the beekeeping century.

THE WORLD SHRANK in the nineteenth century. New inventions in America soon made their way to the Old World, and other innovations hurried back across the Atlantic. Industrialization brought affordable goods to the people; mass-produced books filled homes and libraries, spreading knowledge like pollen. Entomology grew as a subject in the United States, just as it did in Europe. Thomas Jefferson, in his *Notes on the State of Virginia*, had called on Americans to become acquainted with the flora and fauna of their country, a challenge taken up by insect hunters when they went into the field armed with nets and collecting baskets.

In Philadelphia, a young boy named Lorenzo Langstroth, born in 1810, grew up so fascinated by insects that he wore out the knees of his trousers studying ants on the ground. He put down crumbs, pieces of meat, and dead flies to attract insects so he could watch them at close quarters, and roved around one of the city's parks observing the metamorphoses of cicadas. A teacher chastised the six-year-old for devoting too much time to trapping flies in paper cages; when she tore up one such homemade prison, releasing its captives, the boy cried himself to sleep in the dark cupboard where she had sent him as a punishment.

Langstroth left behind bugs to study at Yale, graduating with distinction in 1831; he was later ordained and took on a ministry in Andover, Massachusetts. It was only in his late twenties that his childhood passion for insects was reignited. In the summer of 1838, he encountered a large glass sphere full of honey on a table in a friend's parlor (such jars were attached to some hives for the bees to build their combs in them). This beautiful sight led Langstroth upstairs to the attic room where his friend's bees were kept; "the enthusiasms of my boyish days seemed, like a pent-up fire, to burst out into full flame," he later recalled. Immediately—on his way home—he bought two colonies of bees.

Langstroth began to study bees in earnest. He remembered sitting on his father's knee as he listened to him read Virgil's *Georgics, Book Four;* now he absorbed the works of other such classic writers as Swammerdam and Huber, and also became a devotee of the British Dr. Edward Bevan. In 1848, Langstroth moved back to Philadelphia and started up a larger-scale apiary. It was here that he was to make a discovery that would revolutionize beekeeping all over the world.

Most American bees were kept in simple, hollowed-out logs or plain box hives. All such designs shared the skep's fatal flaw: to remove the honeycomb, you had to cut it away from the surrounding walls. The bees would fill any space between the comb and hive wall, either with more comb or, in the case of narrower gaps, with propolis, the sticky resin gathered from trees.

American bees had at that time been stricken by the wax moth, whose larvae destroy the honeybee's comb and brood. Probably brought over from Europe at the start of the century, the moth had devastated many hives; in 1808, it was estimated that four-fifths of the colonies in the Boston area had been abandoned because of it. This was another issue that a better hive could address.

It was Langstroth's refinement of the basic box hive that was to confront both problems. His influence was to extend far beyond the United States: some three-quarters of the world's hives today incorporate the discovery he was about to make. His simple deduction would change everything.

Langstroth had experimented on hives with frames of comb

attached to top bars that slotted into the hive's body. He left a slight gap between the bars and the hive cover, making the combs theoretically easier to move; but the bees insisted on attaching the frames of comb to the sides of the hive. On October 30, 1851, Langstroth suddenly realized that this gap had to surround the entire frame. He saw that the bees instinctively left a corridor, between and around the combs, that only *just* allowed two bees to pass each other; they filled in anything larger, as bees hate both drafts and wasted space. This corridor, around $\frac{3}{8}$ inch wide, has come to be known as the bee space. If he left this exact distance around the frames of comb, they could be removed easily—both to harvest the honey and to examine the comb for diseases such as the wax moth. Langstroth's discovery was a "eureka" moment; at that instant, he wanted to run down the streets like Archimedes.

Tantalizingly, it was too late in the beekeeping season to put the new principle into practice. All the same, Langstroth applied for a patent for his new hive in January 1852, quit his job at a school for young ladies, and made one hundred hives incorporating the bee

Langstroth's moveable-frame hive: a revolution in beekeeping.

space. This was the first moveable-frame hive: the piece of equipment that is the basis of modern beekeeping.

Langstroth's patent defined the bee space, giving him—at least in theory—rights over every hive in which his modification was used. The discovery had major commercial potential. Improvements to the old box hives, through greater control over the combs, would result in healthier colonies and higher yields. "You have made not just a discovery but a revolution," a friend told him.

It was at this time of high excitement that Langstroth's mental health collapsed. He had long suffered, intermittently, from severe depression—"head trouble," as he called it. Now his problem struck again, in force. Leaving his wife and children in Philadelphia, he took refuge with his brother-in-law. Here, he managed to write his classic book *The Hive and the Honey-Bee* (1853), but was much debilitated. Trying to describe his severe sufferings, Langstroth quoted the seventeenth-century cleric and poet George Herbert:

> *My thoughts are all a case of knives*
> *Wounding my heart*
> *With scatter'd smart.*

In a particularly painful manifestation of his illness, what formerly gave him the most pleasure—his bees—instead caused the most pain, stinging his mind with "scatter'd smart." He would sit on the other side of the house from the hives, hide his bee books, and even, when escaping into other literature, find the capital letter *B* painful because it reminded him of the insect. Relief from such profound melancholy was only to be found in the impersonal field of the chessboard: he would lie awake at night, moving his mind through chess problems.

Instead of bringing Langstroth riches, the patented invention brought strife and misery: this was a classic example of an unworldly inventor coming into conflict with hard-faced business-men. An idea as good as this was bound to spread like wildfire—and it did, irrespective of the rights of its inventor.

At first Langstroth, suffering from his "head trouble" and having little business acumen, tried to get others to capitalize on the invention on his behalf. Meanwhile, others with a sharper approach just took the idea and ran, often making hives of slightly different designs, but still incorporating the concept of the bee space. Langstroth teamed up with a businessman named R. C. Otis, who bought the patent right for the moveable-frame hive in the western states and territories. In defense of the patent right, Otis and Langstroth geared up for a lawsuit against an alleged infringer, Homer King. This New York–based businessman had contracted in 1867 to pay a royalty to Langstroth; he reneged on this three years later, saying he had changed the design and no longer needed to pay. The battle lines were drawn for a court case.

Meanwhile, American beekeeping was burgeoning, and organizations began to proliferate. In 1870, the first national convention of beekeepers, the North American Bee Association, met in Michigan, with Langstroth as president. The following year, the Northeastern Beekeepers' Association grew into a rival national group, the American Beekeepers' Association—and its members also elected Langstroth as their president. The group's leader, Homer King, doubtless trying to put on a good show in front of the beekeeping community, then proposed that a fund of $5,000 be raised to support the bee-space inventor who had been unable to capitalize on his discovery (due, of course, to the maneuvers of men like himself). What was more, promised King, he would lead the fund-raising with a donation of $50. Langstroth's colleague Otis stood up and denounced King, saying the bee master deserved justice, not charity—and he'd give $500 to start a fighting fund. King replied he'd give the inventor $1,000—and draw the check immediately! Otis called for King to be prosecuted. The row made it into the newspapers, which gleefully reported this testy quarrel between bee men.

As is often the case, the row began with a standoff and ended with a whimper. After the deaths of his wife and his backer Otis, Langstroth, still in poor health, felt unable to press the lawsuit; it was dropped, and King and many others continued to make their new hives without paying him a cent. In his *History of American Beekeeping* (1938), Frank Pellett remarks that it would have been better for Langstroth if he had simply given his idea to the world, instead of trying to defend his patent.

Beekeepers managed to unite when the two national organizations merged in 1871 to become the North American Beekeepers' Society. Ladies could join for free; men paid a $1 membership fee. Nobody could speak for more than five minutes at the meetings— apart from Langstroth, who was honored with the right to talk at any time, and for as long as he liked. He had by this time gained the respect of beekeepers who had previously challenged his right to a royalty on each new hive, and was fast being recognized as the father of American beekeeping.

A photograph of Langstroth, aged eighty, shows a kindly face that radiates goodness: at first glance, he has the air of a jolly cleric looking benevolently over his glasses; then you notice a clean innocence to his brow. But the longer you look, the more you perceive a set to his mouth and jaw that is the mark of the survivor, and see a light in his eyes that carries both sorrow and hope. With an unshowy intensity, it is a mesmerizing face; in the photograph, he still appears to be thinking.

Lorenzo Langstroth died in 1895, collapsing in church at the age of eighty-five, and many memoirs and obituaries were published in the specialist bee press. Much was made of his kindness in words and actions. An argument with him (about bees, naturally) would rapidly be followed by a dignified and humane apology; con-

Lorenzo Langstroth, the father
of modern beekeeping.

versations with him were rich with anecdote and learning. "Time always took flight when he became a companion," wrote Albert John Cook, a friend who was a biology professor. Another admirer, the eminent American beekeeper A. I. Root, called Langstroth "one of the most genial, good-natured, benevolent men the world has ever produced." He clearly had his quirks, however. Researchers are still mystified by the secret scribbles in his diaries: there is more to Langstroth than the eulogies would imply.

FURTHER BEEKEEPING innovations followed thick and fast in the second half of the nineteenth century. In 1865, Major Franz von Hruschka, an Austrian living near Venice, gave his son a piece of honeycomb in a basket. The boy swung the basket around his head—a reckless gesture of playfulness—and his father noticed how honey was thrown out of the comb by centrifugal force. This principle was used to create extractors that removed the honey more easily from the comb; previously it had been squeezed out and dripped laboriously through a bag.

Other major advances were made on the age-old practice of using smoke to pacify the bees. You puff away on a pipe and blow the tobacco smoke toward the bees, but this homespun method was not entirely effective. At the first British beekeeping show in Crystal Palace in 1874, a prize was offered for the best "bee-subduing" device. One entry was a briar pipe with a rubber tube to blow the smoke through; another, The Bee Quieter, entered by a Reverend Blight, had a burner attached to a small pair of bellows with a wooden nozzle at the end to direct the smoke. The best design on this principle came from America. Here, one of the foremost proponents of commercial beekeeping, Langstroth's friend A. I. Root, had adapted a tin used for popping corn, filling it with rotten wood and burning coals that he extinguished before himself blowing the smoke into the hives. Root decided, in the end, that he'd rather be stung than smoked out by this Heath Robinson device. Then in 1873, another prominent American beekeeper, Moses Quinby, improved the bellows method so the smoker could be used with one hand. It is this design, with a further refinement patented by T. F. Bingham, that is still used today.

Beekeepers using Langstroth's moveable-frame hives wanted to take honey out of the frames as easily as possible, and to do this, the combs needed to be free of brood. The "queen-excluder" was a perforated sheet, placed above the brood chamber, which kept the queen from going up into the honeycomb and laying eggs because her wider abdomen could not fit through the holes. Primitive versions of this device had been used at the start of the century; improved, mass-produced excluders were common at its end. By now, beekeepers also employed "bee escapes," mechanisms used at harvest time permitting the bees to leave the supers but not reenter.

One of the greatest advances of this time was the introduction of sheets of pre-prepared wax foundation, which formed a base within the moveable frame and gave the bees a head start on building the comb, saving vital energy. Foundation sheets were often reinforced with wire to enable the comb to withstand centrifugal force so the honey could be spun out of the comb in the new extractors.

All these inventions fed off each other, with ideas and designs constantly succeeding one another; in the meantime, humans organized like insects as specialist publications and beekeeping associations spread and strengthened on both sides of the Atlantic. *American Bee Journal* was founded in 1861 by Samuel Wagner, a bank cashier whose passion in life was the honeybee. Wagner had learned German to enable him to translate the writings of the Reverend Johann Dzierzon, a German-speaking Pole working in Silesia who was also working on the concept of moveable frames. When he came across Langstroth, Wagner concentrated instead on promoting his fellow-American's revolutionary idea.

Another prominent journal was *Gleanings in Bee Culture*, published and edited by A. I. Root, who gathered together many practical ideas, some "gleaned" from such specialist publications as *American Bee Journal*, *Bee World*, *Prairie Farmer*, and *Rural New Yorker*. Initially planned as a quarterly, the magazine immediately became a monthly publication after the success of its first edition in January 1873. Root had fallen in love with bees when he first saw a swarm. He questioned everyone he could about his "strange new acquaintances," searched high and low for bee books, and eventually met the famous Langstroth. Root became a beekeeping evangelist, and his journal spread the word. At first, he employed a printing press operated by a treadle. Wind power was added to foot power when he attached a windmill to the machine, not entirely successfully—subscribers unhappy about the printing quality of some of the pages were told the vagaries of the wind were to blame. Root subsequently used the more reliable method of a steam engine to produce *Gleanings*. The subscription list grew year after year: five hundred, then more than eight hundred, and nearly double this by the end of the third year.

≈

THE INVENTIONS being churned out and publicized in the new specialist press were snapped up by beekeepers who, thanks to Langstroth's moveable-frame hives, were now producing large quantities of honey.

This was the start of mass production. The man who did most to build up the honey industry in the United States—the dollar titan of this sweet capitalist gold—was John Harbison, The Bee King of California. Originally from Pennsylvania, Harbison first came west prospecting for real gold in Cavaleras County. He soon turned to growing fruit trees in Sacramento, and it was there that he saw the potential of the honeybee.

By the midcentury, East Coast bees had swarmed naturally up to the mountain barriers, with the exception of those brought to the Salt Lake area by the Mormons in 1848. To bring a large number farther west would be an epic task, and was first undertaken by a botanist named Christopher Shelton. Only one of Shelton's twelve colonies survived, and he himself died when a steamboat he was traveling in caught fire and sank. A certain Mr. Grindley did manage to bring four colonies across the plains in the back of his wagon from Michigan, stopping occasionally to let the bees out for a feed, but it is hard to imagine this method on any scale; other efforts were also only successful in transporting small numbers.

Harbison, though, was a man who thought big. The number of colonies he brought west was soon in the hundreds and his means of travel was by sea.

On his first successful journey, the hives left the family apiary in Pennsylvania on November 15, 1857, to be loaded on board a ship in New York City. The bees were released, once, for a breather, then they journeyed on, crossing the isthmus of Panama by land and continuing by sea to San Francisco—finally reaching Sacramento on yet another boat. In all, the bees had traveled nearly six thousand miles in forty-five days. It was a successful trip, not least because it proved it was possible to bring over a reasonable number of colonies. The original sixty-seven colonies had been reduced to fifty, but these soon multiplied, and there were plenty of people prepared to pay for this rare commodity. The journey cost Harbison around $800; he made more than $12,000 in profit.

Further journeys produced their trials as well as their rewards. Of the one thousand colonies of bees Harbison brought west between October 1858 and April 1859, poor handling meant just two hundred survived by May. Among the six thousand colonies shipped over in the winter of 1859–60 was the dreaded American foulbrood. This bacterial disease, which kills the larvae, began to attack the bees in the West. In a downbeat ending to his account of bringing bees to California, Harbison admits that some lost money, others their reputation: "The result has been bad for all concerned." These early steps and setbacks show the fragile origins of the honey business, surprisingly so, considering how phenomenally successful it was shortly to become.

Californian honey production continued to grow apace. When people heard rumors that Harbison had made $30,000 from his 1859 trading, it sparked something of a bee rush, or "bee-fever" as the headlines put it. The following year, an estimated ten thousand colonies traveled west by this sea-and-isthmus route. By the end of the 1860s, Harbison himself had two thousand colonies, most of them along the Sacramento River, south of Sutterville. During the 1870s, he was the biggest honey producer in the world.

The Langstroth bee-space discovery was by then making an impact, though his patented hive did not solve all of Harbison's problems, as first he'd hoped. Harbison found that Langstroth's patented hive was too small and flat to suit the conditions in California with its magnificent nectar flow. His own major advance was to invent a hive that produced small squares of honeycomb that could be taken straight out of the hive and sold to the public. These 2-pound combs were packed in pails and sold as a package of pure goodness. There began the "honeycomb era." A major load of Californian honey—ten railcars packed full of combs—arrived in New York City from San Diego in 1876, causing an enormous stir.

During the 1870s and 1880s, San Diego County was a region of sage and buckwheat—both excellent honey plants. This was pioneer land; one of the best areas for small apiaries was the backcountry, inland from the coast up to the Volcan, Cuyamaca, and Laguna mountains. Here, smallholders would subsist on a homestead with vegetables and livestock, including a row of beehives.

A gloomy comment came from Harbison, who said pioneer settlement and clearing was destroying the bee industry by damaging its forage plants. A debate ran in the bee press about which plants you could grow to boost nectar supply. There was still plenty to go around, however. More and more citrus and other fruit trees were being planted each year, for example, providing excellent food for the bees. By 1884, Californian apiarists achieved an annual production weighing more than two million pounds. Curiously, given that honeybees help pollinate trees, there was some conflict between fruit growers and beekeepers; Harbison wrote a letter to the *American Bee Journal* in 1893 saying he had lost some 350 hives to arson in just one year due to this. But despite such tussles, California became the leading honey-producing state in the United States.

∼

IN THE LATTER HALF of the nineteenth century, there emerged something of a cult following for what became known as the Italian bee. *Apis mellifera ligustica* is still much admired today for its attractive yellow coloring, productivity, and docility. During the Napoleonic Wars a Captain von Baldenstein served in northern Italy, where he admired these pretty, useful bees. He returned to live in his castle in the Tessin Valley, an area of Switzerland that borders Italy, and later sent men to fetch some of the insects; the bees arrived in September 1843. Johann Dzierzon brought them to Vienna ten years later, and his writings continued to spread enthusiasm for the Italian bee around Europe and over in America.

An early attempt, in 1855, to bring Italian bees across the Atlantic failed because a ship's officer stole some honey and the bees starved. Samuel Wagner, of the *American Bee Journal*, with support from Langstroth, again imported some in 1859, but there were questions about their genetic purity.

There were attempts to get the U.S. government to support the project, with forecasts of spectacular dividends from this fabled honey producer. Then a botanist, S. B. Parsons, undertook the journey himself. He bought some bees from H. C. Hermann, a German beekeeper whose book *The Italian alp-bee, or, the gold mine of husbandry* (1859) had done much to popularize the variety. The bees were packed in cigar boxes filled with honeycomb; in all, twenty colonies were sent off to the States.

Parsons had packaged the bees into three batches; a third for the U.S. government, a third for the beekeeper P. J. Mahan, and a third for himself. On April 18, 1860, when the *Arago* steamer docked in New York, their Italian queens were finally unloaded in the United States—alive, but only just. The combs had loosened in their boxes, and many of the insects were crushed. None survived in the government's batch; none in Mahan's; of all the bees that set out, at a cost of $1,200, only two Italian queens had made it. Once landed, this pair still faced a precarious future. A W. Cart of Coleraine, Massachusetts, took one queen and an Austrian beekeeper, Bodmer, the other. The Austrian failed with his queen; luckily, Cart had more success. He created a large apiary of Italians for Parsons, and these were the genesis of the Italian bee industry in the States. A total of 111 of these queens were taken to California, most arriving in good condition.

ITALIAN QUEENS arrived in England before the United States, imported by Thomas Woodbury (1818–1870). Woodbury was one of the most important British beekeepers of the modern "scientific" era. He studied the great bee writers and, after becoming fascinated by the possibility of other races of *Apis mellifera*, imported one thousand Italian bees from H. C. Hermann. They arrived by train in a box: you can imagine his excitement when, back home near Exeter, he shook the bees out, hunting for the queen. He placed her carefully in a wineglass and carried her to a skep where she was put among other bees. A fortnight later, when he spotted workers carrying plenty of pollen into the hive to feed new brood, Woodbury knew that the colony and its new queen were thriving.

Woodbury was quick to adopt the principle of the bee space and is credited as being the first person in Britain to use moveable frames, in 1860. He also developed his own hive, incorporating Langstroth's innovation. With a typical inventiveness, he was planning a hive suitable for the giant eastern bee *Apis dorsata*, when he died, suddenly, at the age of fifty-two.

Another interesting aspect of Woodbury's life was that he corresponded with Charles Darwin. After formulating his earth-shaking theory of evolution, Darwin continued to live in the quiet seclusion of Down House in Kent, carrying on an extensive correspondence and taking daily walks in his garden where he would peer into the flower beds, minutely observing the behavior of insects. He now consulted pigeon fanciers and beekeepers such as Woodbury about native species. He was most intrigued when Woodbury sent him a sample of the new comb foundation in the 1860s, and also wrote about how bees tend to visit one sort of flower at a time, remarking how important this is to the cross-fertilization of plants of the same species. "Humble [Bumble] and hive-bees are good botanists," he commented, "for they know that varieties may differ widely in the colour of their flowers and yet belong to the same species."

SCIENTIFIC ADVANCES and the development of beekeeping groups led to the growth of a specialist press in Britain as in America. The *British Bee Journal*, launched in 1874, came out monthly, then weekly, and settled down as a fortnightly paper. Events such as the first bee and honey exhibition at Crystal Palace in 1874 helped to spread ideas and inventions. This was set up by the newly formed British Beekeepers' Association, which aimed for "the encouragement, improvement and advancement of bee-culture in the United Kingdom, particularly as a means of bettering the condition of cottagers."

Since the start of the century, beekeeping had become an increasingly useful part of the rural economy. In 1870, a professional gardener named A. Pettigrew wrote his *Handy Book of Bees* (1870), with an attractively plain-speaking tone—not unlike William Cobbett's in the 1820s, but with an even more businesslike perspective. Pettigrew's father had been a laboring man and bee-keeper in Lanarkshire. His sons helped look after the bees, and when the future bee author became an apprentice, then journey-man-gardener in Middlesex, he would keep his own bees in addition to managing his employers' bees. Money was an important incentive: "Stings do not seem half so painful," he wrote, "to the man whose annual proceeds of bee-keeping amount to £10, or £20, or £50." He also praised the occupation for its productive, moral nature and recommended that swarms be given as gifts to deserving servants: "Who has not seen hundreds of working men blessed and charmed beyond description in attending to their bees and cow?" he asked. "Such men are superior to the low vulgarities of the public-house, and superior in every sense to those who waste their time and strength in drinking." He hadn't much time for Italian bees, asserting that a gullible susceptibility to the new was the greatest weakness of an Englishman, and he also found straw skeps infinitely preferable to the newfangled wooden hives.

But Langstroth's advances and a generally more scientific approach to beekeeping were spreading. At the end of the century, a Sussex beekeeper, Samuel Simmins, wrote a book promoting beekeeping as a profitable pursuit. His prediction of the market shows how perspectives had shifted toward industrialization. He talked of how bees, through pollination, helped along fruit being grown for jam, one of many foods that were now mass-produced. Simmins saw that, with the aid of modern extractors, liquid honey was the future now, not comb: "Honey in the comb will ever remain a luxury," he wrote, "but that in the liquid form is destined ere long to be found in general use in almost every family." It was a prediction that has proven true to this day.

CHAPTER NINE

CREATIVE BEE

At the start of the twentieth century, *The Life of the Bee* by the symbolist playwright and essayist Maurice Maeterlinck became a bestseller. This Belgian, who went on to win the Nobel Prize for Literature, wrote his account of the honeybee in an explicitly literary style, which appealed to the general public and not just a beekeeper readership. My copy is a small, moss green hardback, embossed with gold flowers, with fin de siècle tendrils swooping elegantly across the endpapers and spine. The book accompanied me on trains, park benches, and bus seats, slipping in and out of my coat pocket for a month or so. This was the edition's seventeenth printing; I became intrigued by what drew so many people to bees at that time.

The volume had belonged to my great-aunt Isobel and sat on her bookshelves alongside W. B. Yeats, Tennyson, and Rupert Brooke. Bees were a motif in the work of all these poets. Yeats dreamed of a "bee-loud glade" on the Isle of Innisfree; Tennyson dreamt of "the moan of doves in immemorial elms, / And murmuring of innumerable bees." "And is there honey still for tea?" asked Rupert Brooke, a line ever-glowing with nostalgia. Bees represented an old-fashioned idyll as factories churned and cities spread.

Maeterlinck's book starts on a similar note, with the storybook account of his first meeting with an apiarist. This old man lived in the Dutch countryside, a place of little trees marshaled along canal banks, of polished clocks and the musical voice of the perfumed, sunlit bee garden. He had retreated from human affairs, and kept twelve straw skeps painted pink, yellow, and blue to attract the bees. In this beautiful place, Maeterlinck writes, "the hives lent a new meaning to the flowers and the silence, the balm of the air and the rays of the sun."

At times Maeterlinck's descriptions of bees are anthropomorphic. He ponders on the virgin queen's flight, wondering if she is a "voluptuary" who enjoys mating in the air. The big-eyed mating drone is a doomed romantic hero whose unique kiss will lead to his death. The duels between the virgin queens, stinging fights to the death, are vicious as only a dramatist could make them.

But while Maeterlinck waxed lyrical, he also had a background in the craft: he had been a beekeeper for twenty years and kept an observation hive in his study in Paris. His book follows the life of the bee through the life of the year, from spring awakening, to swarming, to the building of a new colony and its filling with honey. He watches events unfold with a sense of awe. Fighting queens will hold back if it looks as if they are about to sting each other to death, leaving the nest leaderless: the central mystery of the hive, for him, was the way the individual bee works for the good of the colony.

Why did they do this? In the pre-Darwinian world, it would have been seen as due to God's divine design. But *The Life of the Bee* is disturbed by the shock waves of Darwin; this was now a world in which humans could be descended from apes. Maeterlinck is tentative about the theory of evolution, perhaps because he saw its implications. His language is still religious in timbre; the mystical life force he sees in a hive of bees is similar to the Holy Spirit. Yet meaning has slipped from "God" to "Life": a mysterious force of nature. The meaning of life, for the bees, is survival. "The God of the bees is the future," he concluded.

~

THE SYMBOLISM of the hive found a novel expression in the building in Paris known as La Ruche (The Beehive). This collection of studios, off passage de Dantzig near Montparnasse, was used by such artists as Chagall, Leger, Modigliani, and Soutine. Originally a wine pavilion for the 1900 Universal Exhibition, La Ruche was designed by the engineer Alexandre Eiffel, famous for his tower. It was dismantled and taken to its current site three years later by the philanthropic sculptor Alfred Boucher. Boucher, a popular artist, spent some of his earnings on a piece of land, where he rebuilt the pavilion to attract artists and writers, leasing out its studios for a small rent.

Boucher explicitly compared his artists' colony to a beehive and called the inhabitants his bees. There were eighty studios in its central, skeplike rotunda. The insects' communal life set an example of artistic productivity: out of the hive comes honey; the studios brought forth sculpture, painting, and literature.

The building is still in use today, after a campaign in 1969 halted plans to turn the site into apartments and a parking garage. The present occupants enter between two jauntily topless female statues, as if at the hive entrance; they pass one another in the hall as they pick up mail, greet one another, look at posters for exhibitions, swap gossip. The floor is covered with brown tiles in a honeycomb pattern and the central stairway goes up the "hive body," to three twelve-sided landings where each wall has a door leading to a studio "cell." The building offers both privacy and company: it is easy to imagine how a "bee" in need of a break would hear a door open, or someone clattering up the stairs, and dart out to meet up on the landing. Artists and writers work largely alone, and yet their antennae need to catch the electricity in the air: La Ruche provides both solitude and social contact, showing how architecture helps people lead their lives better.

~

FIVE MINUTES' WALK from La Ruche is Parc Georges
Brassens, one of those Parisian parks that is calm, civilized, and
well planted. Having been swept out of La Ruche as a gate-crashing
tourist by a classically formidable concierge, I needed to recover
and followed a winding path that led quietly away from the growl-
ing of the city toward trees and flowers. Lines of vines etched one
slope, and a bank of blooms bounced lightly with bees; these signaled the edge of a city apiary with sixteen hives. Paris is a city that
encourages beekeeping; several of its parks have such sites.

Relaxed by this urban idyll, I turned around and suddenly saw
its backdrop: vast, looming tower blocks that were the opposite in
scale and intimacy to La Ruche. These brutish buildings made people and neighboring places insignificant.

Afterward, wandering around the quarter, I was struck by how
many houses from the nineteenth and early twentieth centuries were
decorated with small, natural details: a home adorned with ironwork
marigolds painted orange and green, the gliding metal ferns hinging
a plate-glass door. Even if the countryside was fading to the distance,
people still felt instinctively close to nature, and wanted it in their
urban surroundings, just as they wanted to read Maeterlinck's book
on the honeybee.

~

THE HONEYBEE continued as a positive symbol; but now this
was partly in reaction to the disjunctions of an era of world wars
and grinding industrialization. One radical thinker fascinated by
bees for this reason was the educationalist Austrian Rudolf Steiner.

Steiner (1861–1925) was born into a family that had worked the
land for generations. His father was a gamekeeper on an aristocratic estate, until he became a stationmaster and telegraphist for

Austrian Imperial Railways, just one of many countrymen whose life was shaken out of rhythm by the "march of progress."

The educator thought his own school lessons were overanalytical; for Steiner, the trend toward specialization, which modern science represented, took away the meaning—the spirit—of the whole. "Through the microscope and other instruments we have come to know a great deal," he said in a lecture he gave in 1922. "But it never leads us nearer to the etheric [spiritual] body, only farther from it." Instead of following the mainstream, he developed his concept of anthroposophy from *anthropos* (humankind) and *sophia* (wisdom), a holistic philosophy of education that remains influential today. Having lost the land and all it represented, Steiner wanted us to find our way back to a more natural and "connected" way of living.

Steiner admired honeybees' collective life and used their example to illustrate his ideas about the world. In 1923, he gave a series of lectures about bees to an audience of Swiss construction workers in which he praised the unconscious wisdom of bees, the love in their community, and how each individual bee is part of a whole.

Much to the consternation of a beekeeper who was in the audience, he questioned some of the disruptive practices that had arisen with the new "rational" beekeeping, such as breeding queens artificially in order to improve and disseminate such stock as the fashionable Italian bees. In the past, beekeepers treated their insects in a "personal and proper" manner; now, he pointed out, humans could make profound changes—such as using wooden hives instead of straw skeps—without really considering the effect it would have on these living creatures. We have lost our instinctive knowledge of nature, he said, and this was bound to have major consequences.

∼

STEINER COMPARED the hive to a human being, with the bees circulating like the blood cells in a body. This powerful image was to influence the pioneering avant-garde German artist Joseph Beuys.

Beuys (1921–1986) first read Steiner as a soldier, then as an art student after the war. By the time he died, he had amassed more than 120 volumes of Steiner's writings, around 30 of them scored dark with underlining. He acknowledged the Austrian's influence, writing how he also wanted to sweep away the alienation and distrust people felt toward the spiritual world.

As a young man, Beuys was shot down over the snowy wilderness of the Crimea while serving as a paratrooper in the Second World War. He was found frozen and close to death by nomadic Tartars; as he lay unconscious for eight days, they brought him back to life, wrapping him in felt blankets and salving his wounds with animal fat. This life-or-death experience lay behind the artist's later use of felt and fat in his work; for him, such materials had a metaphorical meaning. Partly inspired by Steiner, Beuys also used honey and beeswax.

For Beuys, there were clear links between bees and creativity: the production of wax, from within the bees' own bodies, was itself a "primary sculptural process." Temperature was as important as space and form in sculpture, he believed, and honey and wax were both natural expressions of warmth. He compared honey and blood, pointing out that they were of a similar temperature (many people comment on the heat of honey fresh from the hive). In an interview in a German beekeeping journal, Beuys spoke of how nectar, "the flower's own form of honey," flowed under the hot sun; he talked also of how beeswax melted to a liquid when heated.

Such conversions represented change; Beuys wanted his art to provoke transformation. There is an urgency in his work—a feeling that it matters—that is lacking in the installations of many of the artists who followed him. He believed in performance art, with a political subtext, rather than permanent creation; it was called social sculpture.

In his 1977 installation *Honey Pump at the Work Place*, Beuys pumped honey in transparent pipes around the Museum Fridericianum in Kassel, Germany. Here, during the hundred days of the artwork, people from all over the world and from many walks of life—economists, community workers, musicians, lawyers, actors, trade unionists—discussed issues such as nuclear energy, urban decay, and human rights. This was Beuys's Free International University, and it was about changing the world: the ideas being discussed should pump through society just as the honey circulated the building.

The meaning of such "actions" relied on the ideas behind them, and the pieces later displayed in museums—photographs, blackboards of scribbles spray-fixed for posterity—are remnants of almost shamanistic events. In this sense, they are similar to the geometric patterns left behind by the honey-hunting rock artists. Beuys wanted to tap into the same source of power as our Stone Age ancestors. In these mechanistic times, he believed animals had the spiritual energy that human society needed; the honeybee was part of the life force we had lost.

THE MANY WAYS in which bees and their communal life have inspired artists and architects are brilliantly explored in *The Beehive Metaphor* by Juan Antonio Ramírez, published in 1998. Ramírez, a Spanish art history professor, has a personal connection to bees; his father, Lucio Ramírez de la Morena, was a dreamer with visions of making money from modern beekeeping. The theory ran that each spring, a colony produced at least one swarm, which could then be hived; an apiary should, in theory, double in size each year, generating endlessly growing profits. In the 1940s, Señor Ramírez began to promote scientific hives by starting a national beekeeping service, converting old-fashioned hives into the modern moveable-frame version, selling them speculatively and instructing beekeepers in the new methods. Alas, his optimism was let down by practice. The venture failed and his hives were impounded.

Meanwhile, Juan Ramírez, who had kept a certain skeptical distance from all these schemes, had taken to reading in the new local library, heading off on his eventual path as an academic. But when he later looked again at one of the hives, he became more interested. He suddenly realized that his father had designed a "building" for bees, with a roof, a window, and commodious spaces in which they could dwell. The parallels between the bee house and human architecture had the power of a metaphor. This thought grew into his remarkable book on how bees have influenced artists in the twentieth century.

$\sim$

ARCHITECTS, the practical artists of society, design buildings for constant, communal use; many have been inspired by the well-designed nests made by social creatures such as honeybees.

This underlying influence can clearly be seen—if you look for it—in the work of the Catalan architect Antonio Gaudí (1852–1926). Gaudí spent long stretches of his childhood in the countryside and later called the pure and pleasing parts of nature his constant mistress. The organic forms of the natural world are obvious in his buildings. One of Gaudí's most distinctive inventions was the parabolic arch, which rises and falls in a single, seamless loop. He disliked the interruptions of traditional columns and arches: this was his sublime solution. What was his inspiration? Ramírez believes it was honeycomb.

The first of Gaudí's parabolic arches was in the bleaching room of a cooperative textile factory in Mataró, built around 1883. He started the commission shortly after the sudden death of his elder brother, Francisco, a scientist whose only published article was about bees; the subject may well have been on Gaudí's mind. The arches that form the body of the building, as Ramírez has pointed out, echo the pendulous curves of wild comb and the hanging chain of bees that starts to build the comb. More explicitly, a drawing Gaudí made for the project replaces builders with bees, and the crest of the cooperative was also a bee, made to his design.

The parabolic arch went on to feature in many of this idiosyncratic architect's buildings, including the Palacio Güell in Barcelona, designed as a home for Gaudí's patron, Eusebio Güell. The main entrance to the building, on a street off the port-end of the Ramblas, is composed of two of these beautiful single-line arches. The central cupola of the building is supported on four further arches and the roof itself is covered in a honeycomb of hexagons, some of which illuminate the dome with geometric stars of daylight. Bee imagery appears elsewhere in Gaudí's work; the Sagrada Familia has a Sacred Heart carved onto the façade, surrounded by the insects, symbolizing how souls are insects sipping from God's nectar in the blood of his son.

OTHER ARCHITECTS working in the first half of the twentieth century were obliquely and explicitly influenced by bees. In 1921, Mies van der Rohe (1886–1969) entered a competition for a high-rise building in Berlin. His submission, called Honeycomb, had the radical idea of using glass for the external walls. If the outside of a building was no longer load-bearing, why not use this structural freedom? This is the modern skyscraper. Juan Ramírez believes Mies van der Rohe's "honeycomb" design has elements of a flat observation hive, which holds a single layer of comb behind two sheets of glass. Traditional architecture hid its engineering; these glass walls showed off the essential structure of both building and honeycomb.

Frank Lloyd Wright (1867–1959) designed buildings that were more obviously influenced by the honeybee. He described his work as analogous to what happened in the natural world: "Building on the earth is as natural for man as it is for other animals, birds, or insects. And as the difference between man and animals grew, so his buildings were converted into what we call architecture." He incorporated the hexagonals found in honeycomb in his buildings from the 1920s onward. The designs for a summer camp building on Lake Tahoe, in western America, is based on six-sided rooms; in the Jiyu Gakuen school in Tokyo, the schoolroom is full of this geometry, down to details such as the backs of chairs; and in his thirties Honeycomb House in Stanford, California, 120-degree angles—like those of comb—replace the conventional right angles not just in walls and windows but also in cushions, a fireplace, and furniture.

It was the honeybee's practice of collective living that most influenced the Swiss architect and town planner Charles Jeanneret, known as Le Corbusier (1887–1965). The son of an enamelist, Le Corbusier enrolled at his local art school during a decade when Art Nouveau was prevalent. But the ideas behind Modernism were starting to stir; he was encouraged to look at the forms underlying nature, and not just their surface, decorative value. One of his earliest designs was a watch case combining a geometric pattern with a bee on a flower. Le Corbusier continued his education in Paris at a time when La Ruche was alive with artistic endeavor. It is more than likely that he knew this famous artists' colony; he later went on to design a building for one of its former inhabitants, his contemporary and compatriot, the avant-garde writer Blaise Cendrars.

Le Corbusier subsequently lived in Berlin at the same time as both Mies van der Rohe and Frank Lloyd Wright. Even more than these two architects, Le Corbusier seems to have been inspired by the internal dynamics of the hive. His vision of the city was a place where many people could live in collective harmony and modernity; he wanted homes, not just businesses, to be in tower blocks, and came up with such concepts as buildings with "precise breathing," which kept the internal temperature at a constant level using methods of insulation similar to those in a hive. His blocks of flats-on-columns echo beehives placed on stands.

AFTER THE SECOND WORLD WAR, Le Corbusier's references to collective living were more muted; communism and fascism had made mass movements more suspect. Juan Ramírez points out how the idea of living in a beehive is a vision of hellish urban living to us today: overcrowded, rootless, and impersonal. In our individualistic, consumer society, we have become increasingly uncomfortable with the notion of close association, such as that of clustering social insects; and meanwhile, the ideals of the tower blocks of Le Corbusier have been discredited through many inner-city failures.

A greater ambivalence crept into the way bees were portrayed artistically. *Spirit of the Beehive*, a masterpiece by the Spanish film director Victor Erice, is set in the 1940s, after the end of the Spanish Civil War, and was shot in 1973, during the last gasp of Franco: the deadened atmosphere of the film conveys two layers of repression, the tense aftermath of war and the stifling fear of a dictatorship. In a bleak Castilian landscape, scoured by wind, Erice shows us a family in which the adults drag through their muted lives, the husband morbidly obsessed by his bees while his wife dreams hopelessly of romantic escape. The torpid frustration of the couple is compared to the alive, imaginative world of their daughter Ana; the inner emotional world of the child offers redemption.

In the film, the bees' world is cruelly impersonal; watching them in the observation hive is no longer a matter of fascination and enlightenment. The man's thoughts about his insects do not lead him out of his despair, only further in. The family's village seethes with distrust, betrayal, and violence; the scurrying insects are no longer a symbol of unity but of the pitiless universe. The honeybee reflects the eye of its beholder; communal life is not a model but a threat.

~

IN THE 1960s, the poet Sylvia Plath incorporated her experiences of keeping bees and meeting beekeepers into her intense form of autobiographical art. This aspect of her work has a further personal layer because her father was a bee expert.

Otto Plath wrote a thesis on bumblebees that was published as a book in 1934. His sweet tooth first led him to bees—as a child, he would follow them to their nests and use a straw to suck out wild honey; he would also catch insects and keep them in cigar boxes to examine them. After growing up in Poland with German parents, he emigrated to live with his grandparents on a farm in Wisconsin. The plan was for him to join a Lutheran ministry. However, the creationist seminary forbade one of his favorite authors, Darwin; Otto went instead into teaching, causing an irrevocable split with his family. He married Sylvia Plath's mother relatively late in life and died, when their daughter was just eight, from complications arising from diabetes. One of Plath's biographers, the poet Anne Stevenson, has pointed out that this condition could have been linked to his sweet tooth.

When Sylvia Plath touched on the traumatic subject of her father, she would sometimes turn to bee imagery. When he died, she said it was as if she had gone into the ground like a hibernating bee. Otto could catch a bee in his hand and put it in his ear, like a trick of natural magic; his daughter wrote of a man who could clench bees in his fist. In 1959, she wrote a poem after seeing her father's grave for the first time. It was called "The Beekeeper's Daughter."

In the summer of 1962, Sylvia Plath and her husband, the poet Ted Hughes, started to keep honeybees. Her diary describes attending a June meeting of beekeepers local to the mid-Devon village where they had a house. Plath was considering a novel based on the lives of the people around her in this secluded part of England: these were notes that could be worked up into literature.

The humor and reportage in Plath's outsider view on village life make a contrast to the dark drama in her poetry. Writing about this beekeeping meeting, she goes into great detail about what each person wears, describing the veiled hats and, particularly, the atmosphere surrounding the rector. At first there is respectful silence around him; then he surprises Plath by tucking her bee veil into her collar—a sudden, curiously intimate moment. The beekeepers, as if licensed by the protection of their white suits, start to tease him about the bees buzzing around his dark trousers (bees are said to dislike such colors and leave lighter ones alone), saying they were his new congregation.

The same diary entry contains a precise, evocative description of what it is like to encounter bees as a novice. After seeing the villagers don their hats, Plath feels increasingly naked, highlighting the vulnerability and anticipation you feel approaching a hive. There is the odd sight of a frame of honeycomb crusted with bees, and the prickles, itches, and tension she feels, now standing in claustrophobic bee clothes in this strange situation.

Plath's sense of unreality is heightened when the Devonians have lost all their homely quirks and have changed, with their garb, into uniform creatures with identical veiled faces. The villagers' unfamiliar actions are oddly ceremonial, and she prays to the spirit of her dead father for protection. The bees fly around as if on pieces of long elastic, a brilliant description of how their free flight is connected to a cohesive whole. Then the meeting ends on a prosaic note, with the secretary selling raffle tickets for a honey show, or as Plath puts it, "chances for a bee-festival."

Charlie Pollard, a local bee man, later brought a box of Italian-hybrid bees to the Hugheses' and the colony was settled in a hive in the orchard, away from the house. When Plath visited the insects, she was delighted to see them entering the hive with pollen on their legs.

~

IN OCTOBER THAT YEAR, Sylvia Plath rose at five A.M. when the sleeping pills she was taking wore off, drank coffee, and began what was to be a series of five poems about bees, written over one week. She was separated from her husband and living with her two small children in a flat in London. Just over four months later, she would be dead.

The experiences in the journal became literature: "The Bee Meeting" expresses her thrumming fear and the way the villagers transformed into "knights" and "surgeons" in their strange garb. She is led toward the hive as if in some sort of initiation rite. "The Arrival of the Bee Box" has the terror of the insect mob; even as she stands amid the bucolic setting of the garden with its pretty cherry trees, she sees power and powerlessness. In "Stings," she writes of the sweetness of the flowers she had painted on the hive; but the mood turns ugly as she thinks of the bees as female: the old, ragged queen, and the drudging workers with their domestic tasks. Another oblique figure—we now know this to be Hughes—is gone. Plath wants to recover, to be a queen, to fly across the sky like a red, soaring comet. The life of the hive is an impersonal engine that has destroyed her, yet she flies in some terrible resurrection. In

"The Swarm," written the next day, she expands the destructive element of bees to the stage of European war; she finds death, power, and defeat within their lives.

Two days later came the last and best poem of the sequence, "Wintering." The honey has been collected, extracted from the combs, and put in jars in the cellar. Plath finds fear in the black of this room; but in the darkness of the hive, the bees are now quietly moving in their slow winter state. The countryside around is pure white with snow. The female bees have gotten rid of the male drones and have entered a time of meditative waiting. There is a quiet, still, serene form of anticipation in the poem, quite different from the excitement Plath felt at the initial bee meeting.

When the bees fly again, they taste the spring. Plath wanted this to be the last poem in her new collection; this hopeful reference to the new year would have been its finishing line.

In the end, the beekeeping poems were published posthumously, as *Ariel*, and reordered, without this final note of optimism. One of the last letters she wrote refers to her plans to resume her beekeeping; it was delivered after her death.

CHAPTER TEN

DISCOVERY

The tradition of the beekeeping cleric continued, following such examples as Charles Butler, Lorenzo Langstroth, and William Cotton, but in England the monastic connection dwindled after the Reformation. Brother Adam, the most famous beekeeping monk of all, however, was based at Buckfast Abbey in Devon for most of the twentieth century; from this quiet base, he made globally important discoveries in his quest to find and breed the best bees in the world.

Brother Adam was born Carl Kehrle in 1898 in south Germany. At that time there was a connection between Germany and the Benedictine Order, who were trying to reestablish a community at Buckfast in south Devon, first founded in 1018. The monks offered the boys they recruited an education and the opportunity to join the community when they were sixteen. In return, they helped out at the monastery. The young Carl was asked by his mother if he wished to go to England "to build an abbey in praise of God." He began a journey, aged twelve, by train, steamship, and horse-drawn taxi to a new life on the edge of the heathy hulk of Dartmoor.

The monks harvested honey as a part of their self-sufficient life, and the frail German lad was thought of less as a builder and more as a suitable assistant to the main beekeeper, Brother Columban. He was soon absorbed in his task. Bees captivate people, Brother

Adam later wrote; he was endlessly fascinated by their precision, order, and ability to adapt to their environment.

Brother Columban had moved with the times by shifting up from straw skeps to the modern frame hives. But there was trouble in the Buckfast apiary, as in the rest of the country. In 1904, reports began in the southeast corner of the Isle of Wight of a disease, probably caused by inbreeding. It blocked the insects' breathing tubes, causing havoc in the colony; it corrupted their wings so they could not fly. Today we know it was caused by the tracheal mite, but then people did not understand its origin: perhaps the new hives were to blame? The problem reached the mainland by 1908, then spread onward.

Many country dwellers a century ago kept bees as part of their cottage economy; the Isle of Wight disease, as it became known, was a rural catastrophe. In the terrible season of 1917, Devon villagers brought their dying colonies to the abbey for help, but to no avail. Of the forty-five Buckfast colonies, only sixteen survived, and these were the ones with Italian-crossed queens. Brother Adam's subsequent work sprang from the idea that breeding could be the key to beating disease.

The British native bee, known as the British Black, suffered most in the outbreak; some think its pure strain was destroyed. This subspecies, in any case, was far from flawless, with a tendency toward testiness and swarming. Brother Adam would spread his cassock around him when he knelt at their hives to stop the irascible creatures crawling up his legs and stinging him. He began to cross his remaining colonies with the imported queens of other breeds. As well as the Italian bees, other kinds were being used by breeders to improve their stock, such as the Carniolan bees, from the Austrian alps, which were famed for their gentleness. At Buckfast, it was the Italian and Carniolan bees that had mated with the native black bee, which fought off the disease. Brother Adam's idea was to build up strong colonies that could develop a natural resistance. His work enabled him to send healthy queens around the country; thanks to him, British beekeepers could restock and recover from this devastating disease, which had killed an estimated 90 percent of their colonies.

Brother Adam was influenced by the ideas of Gregor Mendel (1822–1884), the Austrian monk who discovered the laws of heredity. Mendel had tried to apply his theories to breeding insects, but he knew more about peas than he did about bees. His hives had been kept side-by-side in the old-fashioned bee sheds that are still in use in Germany today; Brother Adam, with his practical apiarian knowledge, knew that the breeds should have been kept separate in order to ensure pure strains. In 1925, he established the isolated Sherberton apiary in a sheltered valley tucked into the high ground of Dartmoor. This collection of hives on poles looked like a forest of postboxes among the granite boulders of the moor. The local beekeepers agreed to keep their hives at a distance when they brought them to the heather in summer.

Brother Adam was trying to combine all the best characteristics: docile bees that built up good colonies, produced plenty of honey, and resisted disease. Artificial insemination techniques would help his efforts. In *The Monk and the Honeybee*, the BBC television program of 1988 that widely publicized Brother Adam and his work, Peter Donovan (his right-hand man for thirty years) picks up a drone and squeezes semen from his abdomen. He draws the fluid up into a pipette. The queen is blown into a tube so her abdomen pokes out of the open end. Two tiny hooks pull open her egg-laying duct. In this unregal position, she receives the semen. She returns to her hive, starts laying, and the controlled breeding program progresses. The queen is crucial to the nature of the colony; she passes her genes onto her offspring and reinforces her dominance by passing out chemical signals to the rest of the hive.

Brother Adam, the beekeeping monk whose extensive travels produced the Buckfast "superbee."

This is why she is the "queen bee." Brother Adam's ambition was to develop the queen of queens.

~

THE SECOND WORLD WAR was not a particularly easy time for a monastery with German connections; to the end of his life, Brother Adam spoke with a strong accent. The white hives on Dartmoor were shot at by patrols and wild rumors spread that they had been placed in a particular formation to guide the Luftwaffe to Plymouth. In one unfortunate encounter with the Dartmoor Home Guard, two of the monks forgot to bring their identity cards to the isolation apiary and were temporarily imprisoned in a pub in nearby Postbridge.

When Brother Adam was diagnosed with heart trouble, he was told he should never work again. But during his convalescence, he rejected all ideas of retirement and decided instead on a series of long and strenuous journeys. His mission was to continue his search for the best races of bees, and he would do so by traveling to collect them himself, be they on mountains or islands, in valleys or deserts. The plan was to incorporate these bees into the Buckfast breeding program. Time was of the essence; modern travel meant that the purity of honeybee strains would not last much longer. In order to get uncrossed stock, he would have to visit some of the most remote places in Europe and Africa.

In 1950, Brother Adam set off in his Austin car on travels that would, over the next twenty-six years, take him more than 120,000 miles. During his travels he went to Provence, following up his interest in the French queens he had already imported; to the Swiss Alps and Germany; to North Africa and the Middle East; to the Mediterranean and others parts of southern Europe, including Greece and the Iberian peninsula.

He made it to the mountainous native territory of the alpine

Carniolan bee, at one point winding around no fewer than seventy-two hairpin bends to reach the bees' isolated fasthold. In North Africa, he was caught up in a sandstorm more vicious and disorientating than any attack by bees. He went to Cyprus, where a rule against importing bees had kept the native strain pure. He went to Crete, fabled in myth as the birthplace of bees; a bee he found here, which had a peppery temperament, was later named after him, *Apis mellifera adami*. He survived a car crash in Turkey, and he went to Mount Athos, a part of Greece controlled by the Greek Orthodox Church, where there were twelve monasteries, scattered hermits, and no females—apart from the bees. On all these trips, queens were sent back to Buckfast to continue the breeding program.

Five varieties of honeybee, in particular, had characteristics he desired. One of the French bees was a good honey producer; a Greek one was good tempered; an Egyptian one, calm. The two best bees were a Saharan bee from Morocco, which proved to be prolific, and a Turkish bee, which was a good honey gatherer and consumed little in the winter, helping to conserve stocks and build up the colony.

From all these trips and all these varieties, the Buckfast honeybee emerged. It became a bestseller in Britain and beyond, especially after tests by the University of Minnesota in 1979 and 1980 proved its superiority to other commercially available queens.

If a bee could survive in damp Dartmoor, it could prosper anywhere. Those who liked the Buckfast breed said the bees were so calm that a colony could be stroked like a pet; those who didn't like the Buckfast said it must be genetically unstable because so many different bees had gone into its breeding. While it is true that the queens need to be produced by artificial insemination in order to keep the stock pure, the Buckfast proved popular with beekeepers all over the world (and continues to be so). The bees even attracted thieves; when some insects were rustled from the monastery in

1982, Brother Adam gave the following description to the police: the victims, he said, were "three-quarters of an inch in length, with dark brown and dark gray stripes."

~

IN 1987, BROTHER ADAM had one last major foreign expedition to undertake. Accompanied by German and British beekeepers, and with the film crew that produced *The Monk and the Honeybee*, he wanted to go up Mount Kilimanjaro in search of the elusive African black bee, *Apis mellifera monticola*.

Did this bee still exist in its pure strain? It had been recorded in the nineteenth century, when it was said to be docile—certainly more so than the fierce yellow African bee, *Apis mellifera scutellata*. The Tanzanians at the bee research station at Arusha were skeptical about the genetic purity of any black bees left. The yellow bees were certainly present, and displaying their traits: when Brother Adam opened one hive, the colony sent out attack signals. Lesley Bill, a member of the expedition and Brother Adam's biographer, reported that the bees were still hovering ready to attack an hour and a half later outside the house where they went to shelter, a quarter of a mile from the hives. Brother Adam said they were the most aggressive bees he'd come across in all his travels.

If the black bee did exist, it would be at an altitude above 8,200 feet; the bees in this remote area were so isolated they would not have needed to develop defense mechanisms to deal with humans and other animal assailants. If they could survive at this height, chances were they could probably survive a northern winter.

The first mountain the group climbed was Mount Meru in Tanzania. Brother Adam, now aged eighty-nine, had been injured in a fall but, undeterred, was carried by two of the beekeepers on the expedition. The sight of the slight monk being borne up the mountain was one of the most memorable images of the film.

Above 8,200 feet, the team started to look in log hives that cooperating local beekeepers had hung from the trees to stop ant and other insect invasions. Unsure of whether the bees would be aggressive or not, they kitted up. Even Brother Adam, who eschewed a bee suit, wore a veil; but he refused gloves, as is often the way with experienced beekeepers, who like to have a more delicate touch on the hives. These bees turned out to be relatively good tempered—perhaps because they contained the prized black bee as well as the more aggressive yellow ones. The queen was captured and put in a balsa wood container, along with a retinue of workers.

The climax of the trip came on the ascent of Mount Kilimanjaro, Africa's highest mountain and a good hunting ground for a pure strain—if one existed—of the African black bee. Brother Adam did not continue all the way up with the rest of the expedition, but the others went on to find more log hives, which they smoked by setting light to a combination of dry grass and elephant dung. In one of the hives, there were numerous crawling black bees, their abdomens banded black on black. This was not, even so, a pure colony. But Brother Adam, when he saw what they had found, was pragmatic: let's put her in the breeding apiary and see what happens, he said. The television program ended with this tantalizing possibility. Sadly, the queens sent back to Buckfast Abbey all arrived dead.

Brother Adam's breeding program was suspended when the monastery decided that its efforts should concentrate on honey production; however, genetic work has resumed and continues today. The beekeeping monk died in 1996; at his funeral, a bell was tolled for each year of his life, and on the ninety-eighth stroke, the coffin was lowered into the ground. His memory is treasured by beekeepers for many reasons. There's the superb heather honey mead he would offer to guests for a midmorning drink, either at the abbey or from a bottle cooled in the stream at the breeding apiary;

there are the stories of his sudden, old-age deafness—quite as selective as his breeding program—in the face of bothersome questions. Above all, Brother Adam is remembered for his pioneering inquiries into different bee races. It was ceaseless work; he toiled as hard as his insects. He popularized ideas about honeybee breeding among beekeepers and the public at large, through his books, talks, and broadcasting. "Everyone is familiar with the guiding principle of St Benedict—*ora et labora* [pray and work]," he wrote. "But those who know his writings better will soon see that a further obligation derives from this teaching, namely that of passing on to others the experience gained in one's life and work."

WHILE BROTHER ADAM was shuffling the honeybee's genes, a brilliant Austrian zoologist, Karl von Frisch (1886–1982), was starting to unravel the mystery of how they communicate. The idea that they can do this at all still has revolutionary implications: the fact that bees pass on information shows they have some form of intelligence. A bee may have a minuscule brain, but its behavior demonstrates significant cognitive ability; in short, it "thinks." Von Frisch's work was so outstanding, so groundbreaking, that it caused an enormous stir—many didn't believe that a "mere" insect could be intelligent. He had discovered that the honeybee's powers of learning and communication rival those of mammals.

This scientist's elegant work is accessible to the layperson; his experiments can be understood by the general reader interested in understanding the observable world. Instead of the higher mysteries of the laboratory, these involved colored cards, boxes imbued with the scent of Italian oranges, and watching bees fly to flowers.

To begin with, von Frisch wanted to see whether a honeybee could distinguish between plants of different colors. At the time, many scientists did not believe invertebrates were capable of this feat. He believed otherwise. To prove this, von Frisch put a small glass dish of honey on a blue card to attract the bees. When he replaced the honeyed card with a blue, unsweetened one, the bees continued to visit: they seemed to be able to see the color and connect it with a reward. When he tried putting shades of gray alongside the blue, to find out if the bees saw tone rather than color, they continued to fly to the blue. The experiment was repeated successfully with different hues.

There was one notable exception. The insects seemed to be color-blind to red. This is interesting because native scarlet flowers are relatively rare in Europe; this is connected to the fact that many of the insects they coevolved with—with the exception of butterflies—cannot distinguish red; some of our native red flowers actually contain a certain amount of purple, a color that *can* be perceived by bees. In addition, other red flowers, such as poppies, reflect ultraviolet light which the bees can also see. The many scarlet flowers native to continents such as North America and Africa evolved to be pollinated by birds and beetles, not by bees.

How keen was a honeybee's sense of color? Von Frisch then put squares of different colors all together and found the bees confused blue with violet and purple. The bees he trained to fly to yellow also went to orange and green. Their sense of color was clearly different from ours, based on a system of visual receptors most sensitive to ultraviolet, blue, and green.

Although color is a visual beckon to bees, this is not all they see; shape, too, turns out to be important. Von Frisch put sugar solution in a box and pasted a radiating pattern onto the front, training the bees to associate the form with sweetness. He placed it among other patterned boxes and found they could distinguish the right one. He repeated the experiment with solid geometric shapes, but found they were better at recognizing broken patterns, such as those found in nature in the form of petals.

~

VON FRISCH THOUGHT there must be other ways in which bees could distinguish one plant from another, and decided to experiment on their sense of smell. When he first began his trials, it wasn't known whether bees could smell at all. But since flowers have different, specific scents that humans can pick up, it seemed likely that the bees could do so too.

First, the scientist put out rows of cardboard boxes, each with a small hole in the front. One box had a dish of sugar water and, for each experiment, either a few drops of an essential oil or a fragrant flower. This was the "food box." Its position was changed frequently to ensure that the bees returned because of the scent and not because they remembered the location. The bees flew around the boxes, pausing at the holes; they entered only the box scented with the smell they thought meant food.

Next, von Frisch wanted to see if the bees could distinguish between different smells. He trained the bees to fly to a sugar solution scented with an essential oil made of the skins of Italian oranges. Then he put this scent among forty-six others. The greatest number of bees flew to the box containing the food scent; they were also attracted to two others—essences of cedrat and bergamot—which, like the original scent, both came from oranges.

By experimenting with different strengths of essential oils, von Frisch discovered the insect's alertness to different intensities of smell to be roughly the same as a human's; if you wander around a garden, following your nose, you are in some ways reacting more or less as a honeybee.

Bees can taste, too, and this sense has an element of calculation; they suck up only nectar of a sufficient sweetness to be worthwhile in terms of energy. Like that of humans, their sense of taste is less subtle and alert than their sense of smell. There is a good reason for this: scent can alert us to approaching enemies, whereas taste happens, for humans and bees, only when it is far too late to flee from danger.

~

VON FRISCH'S most famous discoveries relate to how the insect orientates and communicates; it was he who cracked the "language" of bees. His explorations on this subject started with a simple observation. When honey was smeared on a piece of paper, he noticed it might take hours or even days for a bee to find it; but once one came, as many as several hundred could arrive soon afterward. It seemed the scout bee had passed on good news to the rest of the hive.

To discover how this had happened, von Frisch carried out a series of revealing experiments. He set up a single-honeycomb observation hive and dabbed the bees with different colors in different parts of the thorax and abdomen in an improvised code so that he could distinguish up to six hundred individuals. When a particular bee came to a sugar dish, he could then follow her to the hive and observe what she did next.

Stationed in front of his hive, von Frisch saw the scout bee make a series of circular movements on the surface of the honeycomb; he called this the round dance. The movement could continue for thirty seconds or more in the same place. Meanwhile, the surrounding bees became excited, touching the dancer bee with their antennae and following her motions. The agitated insects then flew out of the hive; shortly afterward, they turned up at the sugar solution.

The insects also seemed to detect the smell of particular flowers on the dancing bee. In the botanical garden in Munich in mid-July, von Frisch counted seven hundred different kinds of flowers in bloom. In this heaven of nectar, he picked on one small plant with silver foliage, *Helichrysum lanatum DC*, which was present in just

one bed. According to botanists, this flower was not normally visited by honeybees. But once the numbered bees had been fed sugar
syrup in a dish surrounded by this flower, many came specifically
to it. From the hundreds of blooms on offer, this was now the chosen one: a scented message had passed through the hive.

Nectar is often located at the base of a flower; to gather it, the
bee may have to crawl right in, and in the process becomes satu-
rated with the plant's smell. When the bee has collected nectar from
some way off, the flower's scent may have been lost from its body,
but is still carried with the nectar in its honey sac; smell can be
transported from some distance.

Von Frisch noticed more facets of the "language." The sweeter
the nectar source, the more vigorous the dance; the bees were communicating the quality of the haul. He also discovered that there
was not just one movement. Bees passing on information about
food from farther away from the hive performed a different one. In
this second dance, the reporting bee ran a short distance in a
straight line, wagging its abdomen vigorously. It did a semicircle in
one direction, then the wagging motion, and then another semicircle in the other direction; when drawn, this looks like a fat figure
eight with a "waggled" center. Von Frisch had noticed this distinctive dance before; many observers of bees have also commented on
it. He now suspected the movement must be connected to information about food.

The existence and length of the waggle-line turned out to indicate the distance of the hive to the nectar. If a bee had to make a
detour from their "beeline"—over a cliff or around a tree, say—it
computed this into its dance. They could also take account of such
factors as crosswinds. All this was important so that the bees could
know how much honey to eat to sustain them to and from the forage patch.

The movement communicated yet more. The angle of the

*"Bee beards" show
how the colony
works as a whole;
here the insects
cluster around a
queen, usually kept
in a cage in the
person's mouth.*

waggle-line on the face of the vertical comb told the bees in which
direction to fly once they had left the hive; it corresponded to the
angle they must fly between the hive and the sun. In fact he discovered the system operated even when the sky was overcast, since
ultraviolet light can penetrate the clouds. The language of bees
could indicate nectar source, direction, distance, quality, and quantity: not bad for "just" an insect.

Von Frisch was so amazed by his results that at one point he wondered if his coded bees had developed some sort of scientific sense and were behaving not as nature intended, but as the scientist wanted. "The language of bees is truly perfect," he said, "and their method of indicating the direction of food sources is one of the most remarkable mysteries of their social organization."

The discovery that these tiny creatures could perform such complex mental feats was a complete surprise to many, and opened human eyes to the capabilities of the animal world. In 1973, Carl Jung said that he previously believed insects to have merely automatic reflexes, but that his views had been challenged by von Frisch's revelation that bees could tell each other where to find food. "This kind of message is no different in principle from information conveyed by a human being," said Jung.

In an introduction to a book of von Frisch's lectures, the distinguished American zoologist Donald Griffin pointed out that the Austrian scientist's work cleared away some of the "melodrama" attached to bees—the elaborated stories that had been so prevalent in their history: for him, the truth was stranger than fiction.

UP UNTIL THE mid-twentieth century, the honeybee was largely seen in the public imagination as a benevolent, if mysterious, part of nature; this was about to change with the dawn of the "killer bee."

The origins of this frightening-sounding creature were innocuous enough. Just as in the seventeenth century, colonizers brought the northern European honeybee to North America; different races of *Apis mellifera* now circled the globe, as beekeepers such as Brother Adam experimented with cross-breeding to create the best possible bee.

In the 1950s, Brazilian scientists, hearing reports of legendary honey crops from African bees—a bounty of 565 pounds was reported from one South African colony—decided to bring some over to see if they could improve domestic yields. A geneticist named Warwick Kerr made a trip to eastern and southern Africa, gathering bees, including the fearsome *Apis mellifera scutellata*, the African yellow bee that later so bothered Brother Adam. Of the many African queens he sent back home, only one survived of the forty-one from East Africa, and fifty-four survived from the south. Nevertheless, these were enough to unleash a chain of dramatic events.

The queens were put into a breeding program and carefully kept in a eucalyptus wood in the state of São Paulo. Queen traps were put on the entrances to the hives to ensure they didn't escape into the wild. Unfortunately, a visiting beekeeper, intrigued by the new bees, came to have a look. He saw pollen was getting caught in the hive entrances and helpfully removed the traps: the "killer queens" were unleashed.

Humans can fiddle with nature, but then can't control it. Bees from twenty-six hives flew out into the wild, where they bred with European bees, and their progeny spread. The Africanized bees traveled 186 to 310 miles a year; by the 1990s there were an estimated trillion such bees in Latin America, with an average of fifteen colonies per square mile—though in some places scientists discovered more than a hundred in this area. Wherever they went, they displaced the more docile European bees. The invading insects spread on and on, and they were heading toward North America.

The Africanized bees moved so rapidly because of how they had evolved in their native territory. In temperate regions such as Europe, the honeybees need to rear large colonies that produce plenty of honey; in winter, they cluster together and survive on these stores. In tropical climates the bees evolved different solutions to different problems. Instead of the fluctuating temperatures of a European summer and winter, dry and wet seasons signal changes in nectar production. More of the bees' energy is spent on producing brood and frequent swarming, or absconding—leaving the nest site altogether—than in making and storing honey. When European bees swarm, they can fly only a certain distance before running out of fuel; Africanized bees can load up with double the amount of food and will fly up to 100 miles before they have to settle in a new nesting site.

The Africanized bees flew on, and as they did, their reputation grew. An early alarm bell rang in a low-key report in *Bee World*, the journal of Eva Crane's International Bee Research Association. "The Spread of a Fierce African Bee in Brazil" ran the headline in 1964. The article told of incidents of stinging, swarming, and absconding, and recommended that no more be imported. But it was too late.

By now, people were frightened by one particular trait of these invading insects. African bees are relatively aggressive toward intruders due to the larger number of predators, such as ants and honey badgers, on their native continent. They are far more touchy than European bees; within seconds of an alarm, thousands can explode from a hive to beat off a potential danger. In one savage incident, a botany student from Miami University was trapped in a rock crevice after disturbing a nest. He was bombarded by bees. Rescuers were fought off by the insects. Eight thousand stings were later counted in the student's lifeless body, around seven per 0.16 square inch.

The story was a gift to the media. Despite the fact that killer-bee deaths were statistically comparable to fatal lightning strikes, the headlines screamed out the news. In 1978, a B-movie called *The Swarm* was released, starring a vast colony attacking nothing less than the United States. One journalist came up with a commercial angle by marketing "killer bee honey," which sold for nearly $1 an ounce. The headlines crescendoed: these "mean" insects could unleash a "vicious frenzy" of stings. "And now, they are heading your way" boomed the *Philadelphia Inquirer* in 1989.

The killer bees, although dangerous enough on occasion, had nothing like the impact on the public that these headlines threatened. But they greatly affected beekeeping. Instead of increasing honey yields, as originally hoped, the new insects had a devastating effect; they were more interested in swarming and spreading than in building up honey stocks. It transformed day-to-day beekeeping. Bees now had to be kept away from livestock and humans, which was a blow to the traditional, amateur backyard beekeeper, and apiarists approaching such insects had to dress in thicker protective clothing and were advised to work in pairs.

In North America, colonies were taken around the country by migratory beekeepers, sometimes called the last real cowboys, who roved from state to state with their insect "herds." These wandering apiarists took the bees to the southern states for good winter foraging. But such movements meant more European-evolved bees would mate with the Africanized ones and perhaps aid the spread of the killer bees.

An eradication program was introduced in California, with Africanized swarms identified and destroyed. It was largely futile. Beekeepers knew the genie was out of the bottle; all they could do was learn how to live with these new residents. After all, people lived happily enough alongside the variety in Africa. In fact, perhaps beekeepers' greatest fear was how the public would react to the inevitable deadly, if occasional, incident.

When I recently asked some American apiarists about the current state of play, they said the bees had reached Texas but seemed to have stopped at Louisiana, perhaps due to the different environment. They spoke with a touch of trepidation; nobody wanted more lurid headlines. Many think there will be a northern boundary to the bees' spread, since they do not cluster and survive the cold as effectively as European bees. Whether this will prove the case is yet to be seen.

CHAPTER ELEVEN

REDISCOVERY

The bee has been used in healing for at least four thousand years. The oldest reference to its medicinal use is from a Sumerian clay tablet of around 2000 BC, recommending river dust kneaded with water, honey, and oil, probably as a cure for a skin problem. Other apian materials, such as propolis, were also favored by ancient civilizations; the Egyptians, Romans, Greeks, Chinese, Indians, and Arabs all believed in their powers. The Greek physician Hippocrates (c. 460–377 BC), for example, thought honey cleaned, softened, and healed ulcers and sores. Such remedies continued as "folk" medicine even after the advent of the modern scientific age, but in a quieter way. Doctors insist on laboratory results, not just hearsay, and manufactured pills have largely taken over from such homespun natural remedies.

In the 1950s, the world price of honey dropped due to an over-supplied market, and beekeepers turned to other products to supplement their income. Pollen, propolis, royal jelly, and even bee venom began to be supplied to the growing alternative health market. There was a term for this trend: *apitherapy*. The movement grew gradually and since the 1990s it has taken off as people have turned back to traditional medicines.

On my mantelpiece, I keep a little bottle of tincture made from propolis harvested from a beehive. This morning, because my throat had a creeping, scratchy tickle, I put six drops of the dark-brown liquid into a small glass of water, where it fizzed like a demon potion. The slightly almondy flavor is just strange enough to taste effective. I began to take it after hearing so many beekeepers praise its usefulness; they scrape a bit of propolis off a hive to chew whenever there are colds in the air. Propolis is a resin exuded by plants to fill in their own "wounds." It protects the bees' colony, too, seeming to combat disease. The insects gather it to plug gaps in the hive, smoothing the inside to stop insects such as wax moths laying their eggs.

Over the 1970s and 1980s, the Danish naturalist Dr. K. Lund Aagard investigated the benefits of propolis, having cured his infected throat by gargling it mixed with hot water; in France, another researcher, Dr. Rémy Chauvin, concluded that it raised the body's resistance by stimulating the immune system. Dentists, in particular, have turned to propolis. For example, Dr. Philip Wander, a Manchester, England, practitioner, uses propolis on mouth ulcers and to clear infections, heal cuts, and stop pain.

~

NEXT TO MY BOTTLE of tincture is a pot of pollen. When I need to feel fortified, I sprinkle some on my porridge or whiz a spoonful in a blender with a banana, some yogurt, and honey for a smoothie. The small, natural pollen pellets have a slightly earthy taste. Their colors vary with the flowers from which they come; I like to look at them in the pot and imagine their origins.

Bees gather pollen as plant dust and knead it into these little balls, which they carry back to the hive on their hind legs and store in the comb. Sometimes known as bee-bread, this form of pollen is a highly nutritious substance, rich in protein, vitamins, and minerals. It is such a complete food that, in one experiment, carried out by Robert Delperee of the Royal Society of Naturalists of Belgium and France, rats fed on just bee-collected pollen and water remained healthy and fertile for several generations. Honeybees, after all, use it to feed their developing young, helping them to grow strong and healthy. Muhammad Ali, the heavyweight champion boxer, aimed to "sting like a bee"; he ate like one, too, boosting his diet with pollen. Abraham Lincoln liked the honey on his bread to be mixed with pollen. Nowadays, nutritionists claim it also helps both female and male fertility.

There is plenty of anecdotal evidence for the effectiveness of pollen. One story, of an American army officer who escaped from a Japanese prison camp in China in the 1940s, recalls how the officer was found by the local people close to death in the jungle and was fed fruit mixed with the plant dust. The locals also dressed his wounded feet with honey and pollen. All this, he believes, saved his life.

THE TWO STRANGEST PRODUCTS of the beehive are venom and royal jelly. How on earth are these harvested? It is one thing hauling frames of comb out of a hive, quite another milking a bee of its poison, or extracting the tiny quantities of royal jelly found in the queen cells of the hive.

Royal jelly is the milky white gelatinous substance fed to both workers and queen bees as they develop; after three days, the worker's brood diet is changed to pollen and honey; the developing insect in the queen cell, however, continues to be fed entirely on the gel, and this enables her to grow to her magnificent size. The queen may live for several years, while in the summer rush a worker bee dies within a matter of weeks; for this reason, royal jelly is perceived as a longevity supplement, especially in China and Japan. The royal jelly yield of a hive is a quarter ounce, and its collection is highly labor intensive, which is why it is so expensive. French beekeepers in the 1950s pioneered the commercial use of royal jelly by creating artificial queen cells and sucking out minute quantities of jelly using a pump.

Bee venom is used to treat arthritis and other inflammatory conditions such as multiple sclerosis, on the principle that it stimulates the release of the anti-inflammatory hormone cortisone. Tickner Edwards, a Sussex beekeeper writing at the start of the twentieth century, described a "patient" arriving at the home of an old-fashioned apiarist for his regular stinging. The therapy is still in use today, with up to eighty bees used in a single session. The insect is held over the inflamed area with tweezers and gently squeezed until it stings.

Bee venom is also collected so it can be injected by needle rather than sting. The toxic fluid was probably first milked at the end of the nineteenth century by J. Langer at the University of Prague. He would squeeze each insect's abdomen and collect its drop of venom in a capillary tube. He had to use 25,000 bees to provide enough for just one microgram in its purified, crystalline form.

In the 1930s, a German firm called Mack started producing bee venom commercially. Originally, the long-suffering employees were made to wait in front of the hives, carefully picking up each bee as it came out of the entrance, and squeezing it so that it stung a piece of fabric that would absorb the liquid. Mack's inventors then worked out how the bees could be boxed and given a mild electric shock to make them sting a piece of paper in defense; a Czech company in the 1960s improved the system further by making the material so thin that the bees could withdraw their lancets and live to sting again. By such methods, bees could be made to inject the paper ten times in a quarter of an hour.

Those who believe in bee venom say it is a natural remedy that avoids the side effects of chemically produced drugs. The medical establishment is, by and large, more skeptical. Experts in rheumatism say there are other more effective, scientifically tested palliatives. However, the first clinical trial of the effect of bee venom on humans has recently been started at Georgetown University in Washington, D.C. The attitudes of some medics toward apitherapy are starting to change, especially where honey is concerned. Much of the credit for this shift is due to a scientist in New Zealand.

PARTS OF NEW ZEALAND are an agribusiness version of the biblical "land of milk and honey." The cows feed on clover; the clover is pollinated by bees; the farmers collect the milk and the apiarists the honey. New Zealand, while a place of great natural beauty, is not quite as untouched as its image suggests; whole areas have been manipulated by large-scale farming. Swathes of native bush were destroyed by the colonials who made great fortunes from sheep farming in the nineteenth century, and some of this land was later given over to commercial forestry. Driving through the North Island, I could see how the hillsides had been scarred and denuded, or swathed with profitable woodland. On the way to the world's largest mainland gannet colony in Hawke's Bay I saw acres of planted conifers on one side of the track; on the other were feathery bushes that produce white flowers. These turned out to be the native plant behind the current revival of honey as a health product: manuka.

Farmers see manuka as a weed, and let it grow only on land too steep and poor to farm; environmentalists are more keen. It is part of the islands' original ecosystem and once mature, other indigenous trees tend to germinate in its shadow; an area of manuka signals a revival of native bush. Beekeepers used to dislike the plant; its nectar produces honey with such a strong flavor that some would bury their crop rather than attempt to sell it. All this has changed; a pot of manuka can fetch more than three times that of even the best monofloral and multifloral honeys.

Manuka is nowadays used for a number of ailments, from leg ulcers to stomach complaints. Its packaging suggests it is medicinal; so does its almost antiseptic taste—even if you like dark honeys; this is not the sort of pot that tempts you to stick in a spoon for a lick. Having watched the rapid rise in popularity of manuka in health food shops, I wanted to meet the man behind its success.

Dr. Peter Molan, a biochemist at the University of Waikato, has spent twenty years exploring the therapeutic uses of honey. His work has demonstrated and quantified its antibacterial properties. He still needs to pinpoint exactly *why* it works so well, but says that he and his team are "getting close." His research has spearheaded its scientific revival; here, at last, was proof of honey's efficacy.

Dr. Molan came to the door of his house with his fat beagle, Jess, who clearly looks up, successfully, to her soft-hearted master at bun time. After growing up in Wales, Dr. Molan wanted to find work in a sunny, politically stable, uncrowded corner of the globe and ended up in Waikato in the middle of New Zealand's North Island. He has done most of his research here on a shoestring, partly thanks to help from the university's international residents, who have translated papers and helped out for free. Honey does inspire good-will. Is it due to sweet childhood memories of honey sandwiches?

Dr. Molan was investigating the health properties of wine-yeast and milk when a friend who was a keen amateur beekeeper persuaded him to take a look at honey. A 1976 editorial in *Archives of Internal Medicine* had dismissed it to the category of "worthless but harmless substances." Was there more to honey than that? His first task was to search the existing literature, to see what investigations were already rolling, rather than trying to reinvent the wheel.

Long-standing honey folk cures turned out to have a certain amount of scientific backing. Sometimes buried in obscure journals, Dr. Molan found references in more than one hundred published papers that suggested that honey was actively beneficial.

First and foremost, honey is antimicrobial. Rich in sugar, it destroys bacteria partly by osmotic force, and also partly through its acidity; but this is not all. The explanation for honey's further antibacterial properties was discovered in the 1960s. It contains an enzyme, glucose oxidase, which catalyzes a reaction that produces hydrogen peroxide, which kills bacteria. This enzyme is destroyed by heat; honey used for health reasons is best processed traditionally, without the heat used in modern, industrialized methods.

Furthermore, Dr. Molan found out that, around the world, certain honeys were favored for therapeutic use. To give two examples, in India lotus honey is used for eye conditions, and in Sardinia that of the strawberry tree is also regarded as particularly healthful. It seems that phytochemicals in such plants come through into the nectar. Because manuka was considered to be a good New Zealand honey for cuts and abrasions, Dr. Molan decided to test its properties. "I'm a great believer that if something is traditional, then it works," he says. "There may be no rational explanation, but that's because we haven't found it."

IN THE UNIVERSITY'S laboratories, Dr. Molan and his assistants tested manuka's efficacy on petri dishes of agar, a nutrient jelly for growing microbes. When bacteria multiply in it, the substance turns murky white. When a hole is punched in the center, and a honey solution is put into it, the surrounding agar becomes clear, indicating the bacteria have died. Using tests like this, he discovered manuka to be particularly effective on a wide range of bacteria. To give an indication of a honey's antibacterial strength, the scientist had the idea of a UMF grading—a "unique manuka factor": the higher the number, the greater the protection, like the grading of sunscreen. The jellybush honeys from Australia are also being investigated; these plants are from the same *Leptospermum* species as manuka.

The long-standing practice of packing honey into wound dressings that had fallen from favor is undergoing a revival. Dr. Molan
successfully tested a DIY version on his wife's boil using a makeup
removal pad daubed with honey; now there are clinical products
using a honey gel.

Honey protects wounds and sores from infection while soothing
the area with its anti-inflammatory properties. It does not stick to
the flesh, so a dressing can be easily removed. Another advantage is
its sweet smell—a bonus when dealing with stinking skin conditions. Honey-dressed wounds have been shown to heal more
quickly and with less scarring than those treated by other methods.
In some cases, ulcers and sores that have suppurated and festered
for months have cleared up in a matter of weeks. In both laboratory
tests and clinical cases, honey has even been shown to be effective
against MRSA, the antibiotic-resistant "superbug" that is of such
concern today.

With scientific proof now supporting them, honey dressings are
more widely available and are undergoing further trials in hospi-
tals of several countries, including Great Britain, the United States,
and South Africa. When Dr. Molan presented all the evidence for
honey to seven hundred specialist wound-care nurses at a conference in Australia, he won a standing ovation from professionals
who understood the potential importance of this ancient cure.

~

BEFORE MY DEPARTURE, I asked Peter Molan to introduce me
to other New Zealand honeys. His cupboard was full of treats, some
more butterscotchy than others I had tried; it was a new chapter in
my honey education. Many of them come from plants that evolved
to be pollinated by birds, before the honeybee was introduced by
Europeans in the nineteenth century; to feed the birds, the nectar
can gush out, and beekeepers with hives in the right place can get

a bounty. Golden tawari honey was especially sweet, rata was fragrant with a hint of spearmint, and pohutukawa is a special, white honey from a tree that grows on the coastline. Was it my imagination, or could I detect a little salt in it? All these came from native plants; my sweet tooth had led me into the islands' ecology.

TAKING "A CHEW OF COMB" is a well-known folk remedy for hay fever; the pollens in local honey are supposed to immunize you against those in the surrounding air. Many people mentioned this cure to me; one person even said you had to find a pot from your birthplace (perhaps this is an example of alternative-health one-upmanship). The renewal of interest in local honey has happened for other reasons, too.

Shop shelves are stacked with honeys from all over the world—fair-trade Zambian rain forest, New Zealand clover, Spanish orange blossom, Italian chestnut. All this offers a form of global shopping that makes life more interesting; honey transports well and has an enormous range of flavors to offer. But globalization can also mean fewer, blander brands, which gain dominance because they have economies of scale; they cost less, but they are less distinctive. To counterbalance global anonymity, we also turn to local food and shops.

Specialist shops can distribute for small-scale enterprises, including those in their area; they help the intricate, alternative web of producers to survive. Such operations have received a helping hand from the thriving health and beauty markets. James Hamill, who runs a honey shop in Clapham, south London, is an actor-turned-beekeeper and sculptor. His combination of trades helps him showcase the bee, and he built an observation hive in one wall so that his customers can watch the insects come and go. The apitherapy products have proved popular—from bodybuilders and sporty types seeking the pollen to others wanting the royal jelly, which James laboriously collects by hand himself.

Like all good beekeepers, James is a purist about how his honey is produced, refusing to blast it with heat, and so keeping its unique properties. He sells monofloral and polyfloral honeys from an eclectic array of sources. When supplies come through, there are honeys from the West Indies, including the fruity variety from bees in mango groves. His own hives are scattered around such places as an old-fashioned orchard in Surrey, on the heather moors of the Isle of Purbeck in Dorset (his all-time favorite is the ling heather honey from here), and in London gardens, where the ever-changing garden flowers provide a rich and varied nectar flow.

James is from a family of beekeepers and was taught the craft from the age of five by his grandfather, who was part of the great Californian apiarian movement, keeping his hives near the orange groves in Irvine. His grandmother had a precious book full of scraps of paper and bits of advice, which was eventually given to her grandson. Every one of the 250 recipes and remedies use honey and other hive products. James has revived some for the shop. There is a cough syrup using honey, glycerine, propolis, and lemon juice— he cannot call it a cure because this simple remedy has not undergone the expensive process of certification—as well as moisturizing cream, lip balms, and a propolis lotion for nappy rash. Not all of James's grandmother's formulas work commercially, such as the rosemary honey shampoo for brunettes, which would need a preservative to have any shelf life, but many are up to the task.

~

AS BUSINESSES GET LARGER, their head offices farther away, and health scares introduce fear to our kitchens, we increasingly want to know where our food comes from. After all, you put food into your body; eating is an intimate activity. This desire for knowledge is behind the success of farmers' markets, where you can meet producers face-to-face, ask questions, and get a sense of who they are, in the process discovering what is on your doorstep.

Local knowledge satisfies another need: nosiness. I discovered two beekeepers working in my part of the world who sold their produce in local shops. Talking to them was like jumping on the back of a bee and flying around my surroundings, over the town and the countryside of the sweeping South Downs, into gardens and buildings and orchards accessible to insects, but not normally to me.

I followed a pot back to its source: Patricia Gilbert, it said on the label, with her telephone number. I rang it. Patricia turned out to be a direct, lively Canadian, now finished with training nurses but unretired in every other sense. She grew up on a farm in Ontario where bees pollinated the clover and honey was spread on the family's homemade bread. When a cousin asked Patricia to help hive a swarm, she realized she was not afraid. Decades later a swarm landed on her fence in Lewes; she called in a local beekeeper, Stephen Kelly. His calm, methodical, easy manner—he is also an elementary school teacher—encouraged her to take up his offer of equipment.

Everyone says beekeeping takes you out of your normal life to focus on the moment. "It is slightly dangerous, so it makes you concentrate," is how Stephen Kelly puts it. Patricia thinks you must be observant. She learned to watch animals acutely as a child; her grandfather would get her to spot which cow in the herd was limping, and her blind grandmother sometimes used her as her eyes. You have to pay attention to bees; Patricia is even careful *how* she looks at them; if the bees are rattled, they will react to the mere flicker of your eye. "You learn to keep your eyes still and use your peripheral vision," she says.

As an amateur pilot, Patricia particularly likes watching the insects in the air. "Their flying is just magnificent," she says. "We have nothing compared to the grace of a bee." She watches them land, flaring slightly, like a plane, to get a cushion of air under their wings to slow them up. They leave the hive, seem to sniff the air, and take off, navigating with their inborn knowledge of flight and air currents. They aren't always precise; she has seen them stumble on the board at the base of the hive when they come in to land, loaded with pollen. But mostly they make us and our machines look clumsy. When the bees swarm, she is fascinated by how they don't bump into one another.

It was a swarm that prompted Patricia to sell her honey. The bees had settled temporarily on a willow tree by a local shop while its owner, Mr. Patel, was outside having a smoke; they got talking about bees; he offered to stock her honey.

People buying Patricia's pots were getting nectar from their own flowers. Lewes is an old town full of established gardens behind high stone walls; the bees soar and sip where they please. But some of her neighbors were nervous, and eventually she stopped housing the insects in her garden. Urban and suburban beekeepers are under pressure; James Hamill received a complaint—in Surrey—that his bees had defecated on some garden furniture. Sometimes too *much* contact with your neighbors is the downside of local life.

But there were other places for Patricia's bees. I visited her allotment, where fellow gardeners, savvy to the benefits of pollination, were generally pro-bee. We stood among pollen-laden hollyhocks and flourishes of yellow fennel flowers, borage and red dessert gooseberries and white currants and apples, leeks going to seed with their big round heads like pompoms on exotic birds. It was a patchwork of plants rather than serried, regimental ranks of other plots and the chairs there made it a place to be, not just to work.

We talked honey. How its use in wound dressings had been common knowledge when Patricia went into nursing in the 1950s. How the bees could fill a hive in a matter of days from a field of oilseed rape. How the local Newick Park Hotel had summoned her hives to their flowering chestnuts, so they could serve their own honey at breakfast. How bees hum at middle C, the pitch rising under threat.

Apparently, swarms can come out of some of Lewes's historic buildings; the house once owned by Anne of Cleves, near Patricia's home, had bees in its old gables. Perhaps they had been there since the sixteenth century, perhaps they'd flown over the head of the hero of the American Revolution, Tom Paine, when he worked as an excise officer in Lewes in the eighteenth century. My thoughts flew back in time with the bees.

Then I went to see Stephen Kelly, the beekeeper who had first helped Patricia. He told me yarns of locals such as Sid Lancaster, who worked the Ouse valley, and whose father took wagonloads by horse to market in Covent Garden. He had three to four hundred hives in the area, and never told anyone where they were; for years people would find old ones abandoned in odd copses around Sussex. "I don't think even Sid knew where half his hives were when he died," said Stephen.

Stephen used to be a bee whisperer, called out to take troubled bees to an isolated apiary in a forest to sort them out. (Sometimes he was more of a bee shouter; people say you must be calm with bees, but when they were in a temper, this steady man found a good shake could startle them into submission.) As an expert, he is often asked to collect swarms; some people now think he should pay for "their" bees; in fact, there tends to be a charge for their removal. Times change. Stephen has seen fluctuations of interest in bee-keeping over the years. There are periods—such as the present— when people take it up, like allotment gardening, to get back to the land. But there were now far fewer beekeepers in Lewes than there once were; in fact, since Patricia left to live in France, I have not found another source of town honey.

Stephen's honey, from the surrounding countryside, is sold by my local greengrocer as well as at Stephen's own door. It is completely different from generic honeys, those blends of whatever is cheapest on the world market, which are flash-heated and micro-filtered to make them stay runny in the pot, unfortunately in the process removing some of the good taste and healthy properties. Small producers tend to leave honey as it is. Stephen simply warms his honey gently to let it run through a coarse filter, removing pieces of wax and so forth. The pot I have of his tastes of Sussex summer. He also sells set honey, a process which occurs naturally, at different rates in different honeys according to the nectar type. Setting can be hastened by "seeding" runny honey with a little of the solid type, which hardens the liquid as the crystals spread.

The rediscovery of local foods is not about pretending to live in a long-gone past, a time when people were more limited to the food produced in their area. I currently have thirty-two honeys on my shelves, sticky columns of pots in different shades of gold. They take me near and far as I travel on a spoon. The exotic and British honeys bring back memories, but Stephen's, in particular, holds such close and constant associations that it will always be a favorite: eating local honey makes your backyard richer.

NOW THAT WE HAVE reached a technological, postindustrial age, there is a reevaluation of the natural world, reflected in the work of artists. The British artist Damien Hirst used honeybees in a television title sequence for short pieces of music by Bach. On a larger scale, the sculptor Robert Bradford made a huge bumblebee at the Cornish environmental visitor attraction, the Eden Project. Backing up the center's theme of biodiversity, the bee climbs a bank of flowers that exists only because of the bee's powers of pollination. While working on his sculpture, Robert became interested in the insects' complex biology, including its ability to communicate and the sexual relationship between bees and blooms. As a figurative artist, he looks to nature to discover different kinds of form and surface—his bee's hairs are made from the nylon filament used to make brooms—and, like anyone who starts to notice bees, he was struck by how we can miss these amazing creatures simply because they are so small in relationship to us: so he made the bee big— about 16 by 26 feet.

The project's buildings contain another large-scale structure that may relate to bees. Eden's vast polymer bubbles are composed of hexagonals, like honeycomb. The architects, Grimshaw and Partners, are part of a movement that is inspired by nature; honeycomb, like the egg, is a prime example of an evolved, efficient form. In a new twist on the "traditional versus modern" debate, such organic shapes are possible thanks to computers capable of accurate specifications for high tech materials and engineering.

THE CANADIAN SCULPTOR Aganetha Dyck has worked with honeybees for the past fourteen years during their short, northern season of July and August. Her fascination has led her right to the heart of the colony. Her work began by placing objects, such as jewelry and even a life-size glass wedding dress, into hives, to explore the sculptural possibilities of beeswax. The bees—her "collaborators"— built comb on them and it was not until the moment of removing the object that Aganetha saw what had happened. The hidden nature of her work, created in the darkness of the hive, was given a further metaphorical layer in *Working in the Dark* (1999–2000). In this work, a poem composed by the poet Di Brandt was put into Braille and placed within the beehive. Beeswax itself is started from a single, anchoring dot; when the fifty-four lines were taken out, the bees had made a new language in this work of "translation."

I asked Aganetha what it was like to work with bees. Like Rudolf Steiner and Joseph Beuys, she responds to their energy; to her, the heat of the hive is part of its mysterious power. After removing a frame, she hovers her hands carefully over the hundreds of bees on the comb, listening to the hum, smelling the scent, and feeling their movement. "It's just the most amazing thing, to have this connection to this warm creature that massages your hands and makes you feel alive," she says. "Especially if you are in the sun and it's very beautiful and the flowers are blooming and you think the world's okay." The warmth of the hive remains in her memory long after she has left the apiary.

She listens, too, first to the sound of the bees to see if they are happy or angry as she approaches a hive. The content sounds are quiet hums, hardly audible until she puts her ear to the hive and hears this gentle breeze of a calm colony. The opposite is the loud and irritable sound of a disturbed hive, when guard bees darting around outside the hive make short bursts of noise—zziiittt zziiittt. Working with her son, Richard Dyck, a multimedia computer artist, Aganetha is recording the noises within. One day they listened to two hives communicating with each other—or at least that was what it sounded like—and heard a low, long moan. Both listened, mystified, to this mass cry of bees. "I am captured by the bees," says Aganetha. "I want to get as close to them as possible." Their size belies their great importance; she is absorbed by "the power of the small."

CHAPTER TWELVE

DO BEES DREAM?

Imagine a Parisienne bee, for once taking the long way home. She flies around the Bastille, up, up, up through cliffs of city buildings; over blue-gray mansard roofs; past ironwork balconies sprouting potted plants; through the narrow medieval streets of the Marais where the sky has been snipped by the buildings into long, slanted strips; over people settling to lunch at boulevard tables. Our bee crosses the river, dips past the poplars on the prow of the Île St. Louis, peers at sunbathers using the quay as a beach, and passes on to the Jardin des Plantes with its artful colors and beds of vegetables. She swings south to the 13th Arrondissement, a district with knots of streets and tree-lined *grands boulevards*, and here she pauses before a small shop in the pretty rue Butte aux Cailles. I stop, too, and enter.

It is Les Abeilles, one of two honey specialists in Paris. A long cabinet lining one side of the room displays the owner's private collection of some two hundred honeys from all over the world. For sale are many French varieties—an intense wild heather honey from the Var; a light, local one from the Bois de Boulogne; rhododendron from the Pyrenees; clover from the Massif Central—and jars from farther afield: almond honey from Spain, mimosa from the Yucatán in Mexico, tupelo from America, pine from Turkey. Bottles of honey lemonade and a fresh-tasting mead sit next to three types of honey sweets doled out with a yellow Perspex spade. Customers come in to stock up on their regular supplies or splurge on a treat. Many are charmed. An old lady pulls in her friend to look at the honey extractor in the window, and reminisce about her youth; honey seems to call up such animated nostalgia.

Jean-Jacques Schakmundes, the shop's owner, is in his sixties, with a proud honey tummy; he refuses sugar, calling it a chemical product, and eats honey instead. A man who changes direction in his work every ten years, for the past decade he has run this shop and a society for city beekeepers, L'Abeille Parisienne (The Parisian Bee). His manifesto for urban bees is clear and passionate. These days, the insects are more protected in the city than in the country-side, he believes. This is because their food is safer. Pesticides can now coat seeds to go into all parts of such plants as sunflower and maize, and beekeepers believe their impact on the bees to be disas-trous. They think the insects' immune systems suffer, their learn-ing abilities decline, and they become disorientated. A lost bee is a dead bee. Many thousands of colonies—billions of bees—have died, mysteriously, in recent years.

In the city, the bees have the pick of the parks and the trees that dapple the light and blossom the spring: acacias, limes, chestnuts, and horse chestnuts; rooftop gardens, flowers in courtyards. The urban, hothouse climate provides an early and long bounty of blooms. If people talk fearfully about stings, Jean-Jacques points out that the risk is very low, plus there is a simple and important equation that matters to most: no bees, no flowers; honey is, he argues, a by-product of pollination.

With the urban bee society, Jean-Jacques set up an apiary of ten hives in the local Parc Kellerman. The group has given the honey to the elderly of the quarter, and to prisoners in one of the suburbs. The site was visited by schoolchildren. Jean-Jacques would ask them why the bees make honey. "To feed us," they'd reply. This is the reality gap, he says. We have a problem until we see that we came after the bees, that we belong to nature, not the other way around. Sometimes people see bees as intruders, he says. "Absolutely wrong. We are the intruders."

Ignorance is one thing; cruelty, another. Last July, the park apiary was torched by an arsonist armed with a Molotov cocktail. All that remained were charred hives and heaps of dead bees—the pathetic devastation of sodden, burnt-out remains.

Despite such animosity, a number of Parisians keep bees in the city, or keep them outside and visit on weekends. "A bee is not a cow," as Jean-Jacques says. "It does not need to be milked every day." There are now some three hundred hives in Paris itself, in gardens, on moored barges, on balconies and rooftops, including the Paris Opera House. The French admiration for the honeybee is reflected in civilized statutes. A Paris law passed in 1895 states that the hives may be 16 feet from your neighbors, except where there is a wall or fence, in which case they can be nearer; and 328 feet from a school. The annual insurance premium for beekeepers all over France is just one euro per hive.

IN THE JARDIN DU LUXEMBOURG, amateurs come to the beekeeping school to take a course that guides them through the year, from feeding the bees in winter to the late-summer collection of the combs. I arrived, on their annual open day at the end of September, to find around thirty people waiting for the doors to open. When they did, a jumble-sale scrum pushed toward a table where the year's crop of pots was on sale. The rush felt exhilarating, rather than unseemly, because this excited sense of harvest is so rare in a city.

The beekeeper, Monsieur Le Baron, had a trim white beard and a steady, grave demeanor. I joined him before the metal-topped hives in the teaching garden, and an informal group gathered, discussing the perils for bees in the twenty-first century. Monsieur Le Baron thought that countryside beekeeping was finished. Why should anyone bother much about the plight of poisoned bees? What were a few beekeepers compared to the might of agribusiness? When we moved on to genetically modified crops, the conversation suddenly became tense. I'd asked about the future of bees, and this was a future that was loaded with uncertainty.

As we were talking, a man from Chicago came up and recounted how he'd kept bees in his youth; when working with his insects, he had felt as if he were a part of their colony. These are happy bees, he said, watching the gold dots, flying without hasty pressure in long, relaxed loops and circles before the park trees in the late-September shadows. We passed ten minutes or so, discussing how the United States Department of Agriculture had classified light honeys above dark honeys, despite their rich flavors; about the gender politics of the hive ("all women love the bit about the drones being expelled," he quipped ruefully); about the eucalyptus honey of California and the blueberry honey of Maine. It was an encounter that was part of the serendipity of the city, and of the subject.

THE CITY is the place where humans gather and hum; the city is where we fly to get the pick of the crop from shops. La Maison du Miel, in the rue Vignon, just north of the haute couture near the Madeleine, is the longest established honey shop in Paris, opened in 1905, with the original mosaic bees still on the floor. The shop started as a cooperative of beekeepers who wanted to get their produce sold in the capital. It is still run by the same family; they now buy other honeys and have seven hundred hives of their own, which they move around the countryside to the best nectar sources. Four pale green drums dispense honey in quantity (acacia, moun-

One of the mosaic bees on the floor of La Maison du Miel *(The House of Honey)*
in Paris.

tain honey, pine, and one other)—some people even heave home
11-pound pails—and there are also forty-five or so other honeys to
choose from, labeled by variety or region.

It is here that I really came to grips with the different character
of honeys and their specific colors, textures, scents, and tastes.
Trying dozens in one sitting was a dizzying task, but various
themes emerged, as I staggered from pot to pot. There were those
with a metallic, herbaceous kick, such as eucalyptus, sage, and mint;
the distinctively fruity ones, like the mango honey that came from
Brazil; the floral lime tree and orange blossom; the colorful sunflower and dandelion; the particular sweetness of carrot honey.
Then there were those that had a special combination of qualities,
for example, the limpid beauty of the acacia honey, with its gentle
taste, emollient texture, and clear glow.

The darker honeys, such as the chestnut, could have the burnt intensity of caramel; the oak honey tasted like wooden fruit, and the pine was a quieter butterscotch. The mimosa had an almost oily texture and a licorice tang. Buckwheat, in this case, was a tang too far for me. *Nappies* (that means "diapers") it says in my tasting notes (though I have tasted it elsewhere with greater pleasure).

The regional honeys conveyed the character of their land, and contrasts were interesting; the honey from the Alps had an acidic kick, while that from the Pyrenees was smoother, with the fruitiness of wild raspberries. If I knew the place, or its stories, this added to my sense of the honey's character; the rosemary honey from Narbonne, which was prized by gourmets of the classical world, made me think of the honey traders of the past.

Printed lists at La Maison du Miel recommend the honeys for particular medical conditions and physical states. Thyme honey is termed a "general antiseptic," which also stimulates digestion; sunflower honey is advised for fever; lavender is said to be good for the respiratory tract and coughs; lime blossom is recommended for sleep and chestnut to accelerate blood circulation. Is this credible? Some of the customers are almost as old as the shop, so it clearly must do *something* for them.

But recently such goodness has flowed less richly. Some of the problems have been due to bad weather. The year 2003 had been disastrous for the honey harvest, explained Monsieur Galland, the shop's current proprietor. There was a drought, which was bad for nectar, then soaring temperatures had meant the bees ate up their honey stores to give them the energy to flap their wings and ventilate the hive. The honey harvest had been down 60 percent in some places; the chestnut trees, for example, bloomed for only ten days. Besides, there was the ongoing problem of pesticides. He was gloomy.

There are still eighty thousand beekeepers in France, but only about 2 or 3 percent of these are professional; and they must get good prices for their honey in order to keep going. Perhaps honey will become, increasingly, a specialized product. Perhaps, one day, we shall look back with astonishment that we took it for granted as a cheap pot to pull off the shelf in any old supermarket.

I ENCOUNTERED some special urban honey in Saint-Denis, just north of Paris. The hives are kept on the roof of the town hall, right next to the gothic Basilique de Saint-Denis. Today a modern observation hive, in a jaunty blue metal casing, had been put next to the church. From the back, it looked like a cross between a public infor-

Olivier Darné's "urban bounty hunter" hive outside the Basilique de Saint-Denis, near Paris.

mation kiosk and a metal Punch-and-Judy stall, with a funnel at the top and bees flying in and out like random smoke. The front had two observation windows, one displaying a comb full on; the other sideways, showing the hive's layers of activity. Some people reclined on deck chairs around it to watch the show, while others peered closely through the hive's windows. The sun reflected gold on the combs, and beyond was the honey-colored stone of the church.

The basilica's doors were suddenly given an impatient rattle. Out jumped Olivier Darné, the beekeeper—or hive installation artist, you might call him—a darting figure in sneakers, with big brown eyes beneath a black-and-yellow skullcap. He settled and flew, settled and flew, jumping up to photograph a coach party peering at his bees, then giving me five pots of successive honeys from this year's harvest, from hives he had placed on the town hall. The honeys progressed through the slightly woody, delicately perfumed freshness of spring, to luscious early summer, to a harsh interlude in July, and then onto rich, mellow late summer. The variation in tastes was astonishing; the early harvests were particularly delicious and complex.

Olivier is a young graphic designer who has kept bees for seven years, first on his roof in Saint-Denis and now in other city spots. The words *Butineur urbain* are written across his blue bee kiosks; "urban bounty hunter" is a rough translation. The honey itself he calls *miel de béton* (concrete honey). Olivier is playful in that French way, with an underlying purpose. With wit and panache, he is bringing bees to people; putting the insects in the heart of the scurrying city; setting the hives among commuters about to dive down into the métro, on their way to gather money at work. His leaflets declared "NOUS SOMMES TOUS DES ABEILLES" (We are all bees). They showed the end of the hive's entrance poking into the sky like the barrel of a tank gun, with bees blown out like living shot. The red-and-white checkered flag flew above, like a standard for battle. Olivier's next plan was to take his hive-kiosks around Europe, to see how a Rome honey differs from a London one. In Saint-Denis, I could see for myself how people were intrigued, delighted, and amused by his bees. This was bringing people to nature, making bees buzz in the urban mind once more.

HOW DO WE SEE nature in the city? Can we dissolve the buildings and the streets that we stand on; can we crack open the hard shell of concrete and tarmac to reveal what lies beneath? How far down is the soil below our feet? Nature feels so distant to our way of operating, even if it is so close. The world moves fast, and so do the bees: can we find the stillness to see them?

What we glimpse of nature in the city is what slips around its edges: the fat autumnal sparrows hopping through the wire around a car park; the early bee stranded on the road. Once your eyes adjust, you can pick out the natural, living world amid the bricks, but you need to focus hard in the urban jungle.

Manhattan, the ultimate cityscape, does not seem a likely home for honeybees. Its surging energy is a force of nature, crackling with the static of action. But these man-made canyons, which so embody human dreams, endeavors, and achievements, feel like no place for bees and blossoms.

Yet when I tried to imagine the naked island as it first appeared to Europeans in the sixteenth century, it began to have distinct possibilities. For a start, the Hudson and East rivers, the very reasons for the city's situation and success, would give the insects plenty of water. The land had been farmed successfully; Harlem was a place of country estates until the subway arrived here in the early twentieth century. The soil of Central Park, and of all the other green spaces, supports the plants that yield the nectar that makes the honey.

~

WHEN THE GREENMARKETS started in New York City in the 1970s, they were making an explicit connection between the food we ate and where it came from. This small eruption of nature in the city encouraged another: urban beekeeping. I arrived at the Wednesday market in Union Square to find David Graves standing at his stall behind pots of his "New York City rooftop" honey, priced at $8 a half pound. Before him flowed streams of shoppers with their flotsam of carts and bags; a dozen different accents made inquiries within an hour, and fifteen children paid a class visit, hustling for tastes from the honey pots. "Do bees like honey?" one asked. "It's their food," David replied, before making them repeat the mantra "gentle honeybees, gentle honeybees." He constantly reassures people that they are safe with these insects in the city.

David first kept hives on a rooftop in his home state of Massachusetts in order to get the hives away from marauding bears. When these mammals emerge hungry from hibernation in the spring, they sniff out the protein of the brood comb and can knock over hives to rip out the contents with an unstoppable greed. Even when David moved his bees up onto the roofs of some outbuildings, one bear still managed to climb up via a birch tree and get its dinner.

An early rooftop adventure went badly wrong. As a novice beekeeper, he put some hives on top of his father's Chevrolet dealership in Williamsburg, Massachusetts. Then he went on holiday. It was a good summer and a tremendous amount of honey built up. The hives became overheated, the wax melted, and one hive collapsed in an overflowing ooze that seeped down through the roof and onto the cars below. His father wanted no more of the bees, so David loaded the hives onto a pickup truck and took them back to his home in Beckett. As he was unloading, one of them tipped over

David Graves, the New York City rooftop apiarist.

and a cartoon cloud of bees chased the hapless beekeeper into the swimming pool. "I've had my ups and downs with bees," he says.

The highs and lows, these days, involve elevators and subways. David first thought of harvesting urban honey as a means of selling a truly local food to New Yorkers; it is a premium product that is low in supply and high in demand. He advertised for rooftop apiary sites by putting a notice on a box of bees on his stall: "We are very gentle, we like to share our New York City honey, do you have a rooftop?" and offered a percentage of the honey to those who took him up. There were takers. Now much more experienced, a Johnny Appleseed of urban bees, David shuttles between the seventeen hives he tends in the city. The highest is twelve stories up, on a hotel in the middle of Manhattan, and there are others on roofs in Brooklyn, the Bronx, the Upper West Side, in midtown on the East Side; on a church, in a community garden, and even by a school. Also, on top of a Harlem soul food restaurant, Amy Ruth's, where the cooks add the honey to their special Southern Fried Chicken recipe.

Apart from being an unusual marketing point, the city is simply a good source of nectar. "When you look at aerial photos of Manhattan, it doesn't seem very green," David says, "but when you go to street level, you start to see there's a lot around." There are crab apples, linden trees, and Bradford pears; the clover, sumac, and tidy flowers surrounding apartment complexes; the plants on roofs and of course the flora of the parks. But there's still too much wasted space; from one of his rooftop viewpoints, David can look into Bronx backyards and muse how much more fruit and vegetables could grow in them as food for humans and bees.

But beekeeping is not exactly encouraged in New York City. The official list of animals banned from the city includes, not unreasonably, bears and large rodents; then there are the likes of "even-toed ungulates," such as deer, giraffe, and hippopotamus, and also "odd-toed ungulates"—other than horses—such as zebra, rhinoceroses, and tapirs. Then the list becomes less consistent, however. There is a section against "all venomous insects, including but not limited to, bee, hornet and wasp." Most stings are likely to come from insects such as wasps and yellow jackets; but the honeybee is a force for the good.

All this might seem like a bit of a technicality, except that, shortly before I visited New York, one of David Graves's rooftop apiarists had just received an officious piece of paper from the city's health department.

~

JILL LAURIE GOODMAN is a lawyer living in an Upper West Side brownstone, a short flight from Columbia University and its spacious campus. Having grown up in a family of gardeners, she likes to keep one foot in the soil, and was immediately attracted to David's advertisement for hive sites. She went home expecting her family to shoot the idea down in flames. Her husband, Melvin, explained to me that, since he was terrified of insects—or at least, like Woody Allen, at two with nature—the whole idea was ludicrous: it had to be done. So they gave their house keys to a complete stranger and invited sixty thousand insects to live on their roof.

The first year, David kept his bees on Jill's roof, and she followed him around learning the craft. The next year, she nailed together a hive and received a box of bees by post. They arrived by priority mail in a small wire cage with a notice declaring them "gentle honeybees." Apparently, the New York postal service can be less than swift when delivering bothersome packages that require signatures; but, for some strange reason, the bees arrived without delay. The queen is in a small, internal cage, with a few workers and a candy plug that she eventually eats her way through. By this time, the workers are familiar with her presence, and together they form the nucleus of the new community.

The rooftop bees take Jill to another world. "There is a magic in producing the food you eat," she says. "Working with bees you have to be totally patient. Whatever tension or anger you have from the rest of the world, you just have to let go." I sat in Jill and Melvin's kitchen eating some of the year's crop, a pale, golden honey, while this intellectual Jewish couple discussed how the rabbis get around the dietary laws saying that you can't eat insects or produce from an unclean animal; how honey is part of traditional ceremonies, such as the bread and apples dipped in honey and served for a "sweet and good" new year; and how children used to be given letters dotted with honey when they first went to school, to associate learning with sweetness.

In the city, bees are kept on rooftops so that the flight path in front of each hive isn't disrupted by humans—one reason for stings. This way, they are likely to be even safer. The New York City health department, however, does not see it this way. It was after four years of keeping bees that the city council's notice arrived at Jill's door. She didn't want it; she wanted the whole thing to go away. Yet it was here. It warned her of a "large number of flying bees found on the roof area causing nuisance." It accused her of "harboring bee hives on the roof of an attached building."

In her fight to keep her insects, Jill discovered that the city's ordinance on animals might allow for waivers. She also made connections with other urban beekeepers. In Chicago, the mayor Richard Daley was so "green" that he had planted the roof of the city hall with 20,000 plants, of more than 150 species, and put 3 beehives up among them. It was San Francisco's official policy to increase the production of home-produced food, including honey. But New York City, it seemed, was beephobic.

~

APIARISTS ASSOCIATE, as do their insects; put any bee-related subject into an Internet search engine and you find yourself linked into a highly active and wide-ranging human network. You can even hunt out a website on bee beards, in which a queen bee is put in a small cage on a human, and the swarm follows to create a calm, living "beard." Busy discussions continue in the real world as well as the virtual one. I encountered one such meeting in upstate New York. The Empire State beekeepers were holding their annual conference in a hotel in out-of-season Alexandria Bay, on the banks of the St. Lawrence River. I joined every sort of beekeeper there, from industrialists with hundreds of hives to backyard hobbyists. They came to chew the fat, listen to politicians make endless promises, and to hear talks on such topics as profitable queen rearing and primitive candle making in the Dominican Republic.

In the evening, we stayed awake for a historical lecture, dozy after the eat-everything buffet. A stray person, thousands of miles from home, I was feeling a little existential around the edges, touched by the chilled waters and tousled skies outside the Bonnie Castle Resort, as the speaker told of Cupid's bowstring of bees and the sweet pain of love. Some of the conference had depressed me. During the past two days, one subject had cropped up with an ominous regularity: chemicals. Honey is still—rightly—regarded as a pure food; but there are threats to its wholesome reputation.

Around the world, beekeepers have had to deal with the terrible plague of the varroa mite. Originally *Varroa jacobsoni*, now called *Varroa destructor*, this small red mite, just visible to the naked eye, happily coexists with the Eastern honeybee *Apis cerana*. But when it crossed over to *Apis mellifera* honeybees, first in the former Soviet Union, the mite began its devastation. It lays its eggs on the brood comb and the larvae hatch so deformed as to be useless. The whole colony is weakened, becoming more susceptible to viruses and other diseases; eventually, it collapses. The mite spread to the rest of the world from the 1960s onward; even the proudly isolated New Zealand got caught out, after a budget cut removed checks on bee colonies around ports and airports.

There are other diseases, too, particularly American foul brood and European foul brood, not to mention the troublesome hive beetle in America. In each case, the problem spreads easily, through the comings and goings of trade. When the varroa mite was first discovered in the United States, in Wisconsin in 1987, it was traced back to Florida, probably to hives kept in the vicinity of Orlando airport; a freight worker reported bees escaping from the holds of international airplanes. Since half of the state's colonies were moved around to pollinate plants in up to twenty other states, from almond trees in California to blueberries in Maine, it is not hard to see how the mite spread. In the year after its discovery, 90 percent of Florida's colonies were destroyed by the disease.

How to combat the rampant bee diseases of the modern industrial world? For a while, products such as Apistan have held varroa at bay. It is meant to be applied in the winter, before the bees are actively producing honey, and does not dissolve in the honey. In theory, as with all such products, the end food is safe. But can you always trust food producers? There have been enough scares to make us wonder. Questions have recently been raised after America and the European Union banned the import of Chinese honey when residues of an antibiotic, chloramphenicol, used to combat foulbrood disease, were found in a number of tested samples. Since China provided a large amount of the world's honey, cheap, generic potfuls—the sort labeled "produce of more than one country"— could well have found their way onto many a breakfast table. The headlines, this time, were measured; the risk to human health was mainly to those susceptible to a rare but serious blood disorder. Anyone interested in the safety of our food took note, however.

Even the terrible advent of varroa has begun to look like "the good old days": the mite has started to develop a resistance to Apistan. What is the next step? The question loomed over the Empire State Beekeepers' conference. There are natural methods using herbs. Others are experimenting with formic acid, which was discovered when scientists noticed birds rubbing their infested feathers with ants, producing acid to get rid of mites. Meanwhile, some beekeepers are breeding queens from Russian stock that may have a greater resistance to varroa. The honeybee genome has now been sequenced, and in the future genetic engineering could speed up the process of breeding for resistance.

Others are trying to go back to the old, pure races. Kangaroo Island, off the coast of South Australia, has become famous for its Italian bees. Introduced in the 1880s, an act of Parliament shortly afterward forbade the importation of any other kinds of bees, and the isolated island is now a haven for what may be the last pure stock of this prized race and an important genetic resource.

There are those who think the industrialization of the honeybee is behind its current problems, that we have put too much pressure on its highly evolved systems. In Rudolf Steiner's 1920s bee lectures, he warned that the artificial breeding of queens could have dire effects. When a beekeeper in the audience objected, Steiner replied that they should talk again in a hundred years' time. That time is nearly up. Queen rearing has transformed beekeeping—people need not worry about losing their stock through swarming and can manipulate the nature of their hives simply by buying new queens—but perhaps some of Steiner's concerns still need to be addressed; perhaps we are pushing bees too far out of their natural behavior.

At the conference, I was disturbed to discover that some beekeepers were starting to put the deadly organophosphate, coumaphos, into their hives, to kill the varroa mite and the hive beetle. Such chemicals work by disrupting the nervous system: this is serious stuff. Although all such products must go through stringent controls, it is not hard to imagine how a less scrupulous beekeeper, pushed against his margins, trying to cure his bees and produce a crop, might treat chemicals in a more casual manner. The specter of China's problems might not be enough to stop independently minded individuals—as beekeepers so often are—from polluting their hives and perhaps putting themselves at risk, as well as honey's pure reputation.

Everywhere this mite has gone, beekeepers have been forced to change their customs, and many have stopped beekeeping altogether. I spoke to one old hand at the conference who kept just a few hives now, and had a philosophy of respecting his insects: "If you keep bees, you have to learn from bees," he said. "I'm seventy-seven and I'm still learning." This man gave a wide berth to organophosphates, and was hoping to find bees bred with a tolerance to the varroa mite. He pointed out another sad aspect of the varroa epidemic: once, many people were "bee*havers*," rather than bee*keepers*, with a few hives in their backyard, he said. Diseases had made this impossible; they required too much intervention; gradually people had given up, or not replaced their dead colonies. It seemed such a shame that bees were moving out of neighborhoods: how would we keep in touch with the honeybee if it moved away from us?

WHEN A NEIGHBORHOOD'S honeybees depart, people might not miss the bees much, but they often notice their gardens producing fewer vegetables; they miss the bees' powers of pollination. About four-fifths of the world's plants rely on pollination by animals, mostly insects; a third of the food we eat comes from plants that exist thanks to them.

But there are problems here, too, and sometimes from an unlikely source. In a book published in 1996, *The Forgotten Pollinators*, the authors Stephen Buchmann and Gary Paul Nabhan point out that the spread and success of the highly successful hive honeybee, *Apis mellifera*, has encroached on the territory of other kinds of bees, with a consequent impact on biodiversity. Some species now exist in fragile "islands" that are in danger of sinking beneath the swelling sea of sameness.

Our casualness toward bees of all sorts is all the more remarkable because, even as they suffer from pesticides and impoverished ecosystems, scientists continue to find them a constant source of fascination. They are currently being investigated as potential scouts for land mines; for a deeper understanding of social evolution; for their ability to "talk" in an age of mass communication. The International Bee Research Association, based in Cardiff, Wales, has a library of 60,000 papers, 4,000 books, and 130 journals; they produce a quarterly publication of 350 apicultural abstracts, gathering the latest research from around the world.

There is much still to discover. As we've started to think of ourselves more as animals, we can now believe our fellow creatures to be capable of greater feats. What about animal consciousness—do bees dream? If we see ourselves as part of nature, rather than above it, we can explore its parts—not least the honeybee—with a renewed sense of awe.

> *But, for the point of wisdom, I would choose*
> *To know the mind that stirs between the wings*
> *Of bees . . .*
> —George Eliot, *The Spanish Gypsy*, 1868

I SOMETIMES THINK of a place that first led me to think about honey. The garden of the late filmmaker Derek Jarman feels like the end of the earth, with its shingle spreading to a seaside view of the nuclear power station at Dungeness, England. One summer, I lay in his garden with my eyes closed and smelled the salt tang of the sea and the scents of the flowers, and listened to the bees among his architectural plants. One of his diary entries mentions how he watched the bees crawl hungrily up the green woodsage. I found a local beekeeper, Malcolm Finn, who harvested this clear, fragrant woodsage honey, and sold it through a roadside stall. Otherwise, he serviced Coca-Cola vending machines in China. All this planted the taste of a place in my mind, and made me realize the connections between humans, insects, and plants.

While writing this book, I had many such fly-by encounters where information and then further phone calls, e-mails, and post-cards were exchanged. Everyone I met, or knew, had some apian anecdote to divulge. I heard a tale of a mysterious wood in Romania, where Romany gypsy bands went for an annual trip on hallucinogenic honey; a friend spoke of his university mate who had a large bee tattooed on his arm, symbolizing how his departed girlfriend had pricked the bubble of his illusion, a reference to the Paul Valery poem *"L'Abeille."* Someone else offered the anecdote of how a bees' comb had once dropped into the Queen Mother's soup. Yet another mentioned how propolis had been used in varnishes in Italy from the sixteenth to eighteenth centuries; its presence might be one of the secrets of the tone of Stradivarius's violins, perhaps because of the quality of the propolis in Cremona, which came from poplar trees. A couple brought back a pot of honey for me from the Karoo in South Africa, a special place with its beautiful fynbos flora, and one of the longest inhabited regions on earth.

Each reference reflected something of the giver's outlook or experience: my diplomat brother would come across bees in flags and local customs; a friend's mother revealed how many years ago she had rushed up to a remote corner of her family home, where the insects had nested, and "told the bees" of her recent engage-ment. The story moved me; she had shared such a special time of happiness and expectation by performing this archaic custom, and it showed me, once again, just how close and important bees have been to humans, even within living memory.

Such precious connections between people, bees, and plants have grown and gathered for millennia. Will they continue if bees withdraw further from our lives? If we lose such closeness, an inti-mate part of our contact with nature falls away; if we lose our respect for these miraculous and mysterious insects, it is at our peril. For life is all one: as big as the world and as small as the honeybee.

BIBLIOGRAPHY

Adam, Brother. *In Search of the Best Strains of Bees*. Hebden Bridge, W. Yorks: Northern Bee Books, 1983.

Adam, Brother. *Bee-keeping at Buckfast Abbey*. Geddington, Northants: British Bee Publications, 1975.

Aldersey-Williams, H. Zoomorphic: *New Animal Architecture*. London: Laurence King Publishing, 2003.

Alexander, P. *Rough Magic: A Biography of Sylvia Plath*. New York: Da Capo Press, 2003.

Allan, M. *Darwin and his Flowers*. London: Faber & Faber, 1977.

Alston, F. *Skeps, their History, Making and Use*. Hebden Bridge, W. Yorks: Northern Bee Books, 1987.

Barrett, P. *The Immigrant Bees 1788 to 1898,* 1995.

Barrett, P. *William Cotton*.

Beuys, J. *Honey is Flowing in All Directions*. Heidelberg: Edition Staeck, 1997.

Bevan, E. *The Honey-bee: Its Natural History, Physiology and Management*. London: Baldwin, Cradock & Joy, 1827.

Bill, L. *For the Love of Bees*. Newton Abbot, Devon: David & Charles, 1989.

Bodenheimer, F.S. *Insects as Human Food*. The Hague: Dr. W. Junk, 1951.

Brothwell, D., Brothwell, P. *Food in Antiquity*. London: Thames & Hudson, 1969.

Brown, R.H. *One Thousand Years of Devon Beekeeping*. Devon BKA, 1975.

Brown, R.H. *Beeswax*. Burrowbridge, Somerset: Bee Books New & Old, 1981.

Brown, R.H. *Great Masters of Beekeeping*. Burrowbridge, Somerset: Bee Books New & Old, 1994.

Buchmann, S., Nabhan, G.P. *The Forgotten Pollinators*. Washington, D.C.: Island Press, 1996.

Budge, E.A.T.W. *The Mummy*. New York: Collier Books, 1972.

Butler, C. *The Feminine Monarchie*. Hebden Bridge, W. Yorks: Northern Bee Books, 1609, facsimile 1985.

Butler, C. *The World of the Honeybee*. London: Collins, 1954.

Campbell, P. *Travels in the Interior Inhabited Parts of North America in the Years 1791 and 1792*. Toronto: Champlain Society, 1793, reprinted 1937.

Charles-Edwards, T., Kelly, F. (eds). *Bechbretha: An Old Irish Law-Tract on Bee-Keeping.* Dublin: Institute for Advanced Studies, 1983.

Cobbett, W. *Cottage Economy.* Oxford: Oxford University Press, 1823, reprinted 1979.

Coggshall, W.L., Morse, R.A. *Beeswax: Production, Harvesting, Processing and Products.* Cheshire, CT: Wicwas Press, 1984.

Crane, E. (ed). *Honey: A Comprehensive Survey.* London: Heinemann, 1975.

Crane, E. *The Archaeology of Beekeeping.* London: Duckworth, 1983.

Crane, E. *The World History of Beekeeping and Honey Hunting.* London: Duckworth, 1999.

Crane, E. *The Rock Art of the Honey Hunters.* Cardiff: International Bee Research Association, 2001.

Crane, E. *Making a Bee-line.* Cardiff: International Bee Research Association, 2003.

Cronin, V. *The Golden Honeycomb.* New York: E.P. Dutton & Co., 1954.

Davidson, J. *Courtesans and Fishcakes.* London: Fontana Press, 1998.

Day-Lewis, C. *The Eclogues and Georgics.* Oxford: Oxford University Press, 1999 edition.

Digbie, Sir Kenelme. *The closet of the eminently learned Sir Kenelme Digbie, Kt., opened...* London, 1669.

Donovan, R.E. *Hunting Wild Bees.* New York: Winchester Press, 1950.

Dummelow, J. *The Wax Chandlers of London.* Chichester, W. Sussex: Phillimore & Co., 1973.

Duncan, J. *An Introduction to Entomology.* Edinburgh: W.H. Lizars, 1843.

Edgell, G.H. *The Beehunter.* Cambridge, MA: Harvard University Press, 1949.

Edwardes, Tickner. *The Lore of the Honey-bee.* London: Methuen, 1908.

Edwardes, Tickner. *The Beemaster of Warrilow.* London: Methuen, 1923.

Erikson, E.H., Carlson, S.D., Garment, M.B. *A Scanning Electron Microscope Atlas of the Honey Bee.* Ames, IA: Iowa State University Press, 1986.

Flower, B., Rosenbaum, E. *The Roman Cookery Book ... by Apicius...* London: Harrap, 1958.

Fraser, H.M. *Beekeeping in Antiquity.* London: University of London Press, 1951.

Fraser, H.M. *History of Beekeeping in Britain.* London: Bee Research Association, 1958.

Free, J.B. *Bees and Mankind.* London: George Allen & Unwin, 1982.

Fry, C.H. *The Bee-eaters.* Calton, Staffs: Poyser, 1983.

Galton, D. *Survey of a Thousand Years of Beekeeping in Russia.* London: Bee Research Association, 1971.

Garlake, P.S. *The Painted Caves.* Harare: Modus, 1987.

Garlake, P.S. *The Hunter's Vision.* London: British Museum Press, 1995.

Gayre, G.R. *Wassail! In Mazers of Mead.* Chichester, W. Sussex: Phillimore & Co., 1948.

Gibbons, E., introduction by John McPhee. *Stalking the Wild Asparagus.* Putney, VT: Hood, 1962.

Gould, J., Gould, C. *The Honey Bee.* New York: Scientific American Library, 1995.

Griffin, D. *Animal Minds.* Chicago: The University of Chicago Press, 1992.

Hauk, G. *Towards Saving the Honey Bee.* Kimberton, PA: Biodynamic Farming and Gardening Association, 2002.

Hauser, M. *Wild Minds: What Animals Really Think.* London: Penguin, 2001.

Hayman, R. *The Death and Life of Sylvia Plath.* Stroud, Glos: Sutton Publishing, 2003.

Hill, J. *The Virtues of Honey.* J. Davis & M. Cooper, 1759.

Hodges, D. *The Pollen Loads of the Honeybee.* London: Bee Research Association, 1952.

Holt, V.M. *Why Not Eat Insects?* Hampton, Middx: E.W. Classey, 1885, reprinted 1967.

Hubbell, S. *A Book of Bees.* New York: Mariner Books, 1988.

Huber, F. *Observations on the Natural History of Bees.* London: Thomas Tegg, 1841.

Langstroth, L.L. *Langstroth on the Hive and the Honey-bee, a Bee Keeper's Manual.* Northampton, MA: Hopkins, Bridgeman & Co, 1853.

Lévi-Strauss, C. *From Honey to Ashes.* London: Jonathan Cape, 1973.

Lewis-Smith, D., Dowson, T. *Images of Power: Understanding San Rock Art,* 1989.

Maeterlinck, M. *The Life of the Bee.* London: George Allen & Unwin, 1935.

Mandeville, B., introduction by Phillip Harth. *The Fable of the Bees.* London: Penguin, 1970 edition.

More, D. *The Bee Book.* New York: Universe Books, 1976.

Munn, P., Jones, R. *Honey and Healing.* Cardiff: International Bee Research Association, 2001.

Naile, F. *The Life of L.L. Langstroth.* Ithaca, NY: Cornell University Press, 1942.

Opie, I., Tatem, M. *A Dictionary of Superstitions*. Oxford: Oxford University Press, 1989.

O'Toole, C., Raw, A. *Bees of the World*. London: Blandford, 1991.

Pellett, F.C. *History of American Beekeeping*. Menasha, WI: Collegiate Press, 1938.

Pellett, F.C. *American Honey Plants*. New York: Orange Judd, 1947.

Pettigrew, A. *The Handy Book of Bees*. Edinburgh: Blackwood, 1870.

Plath, S. *Collected Poems*. London: Faber & Faber, 1981.

Plath, S., Kukil, K. (ed). *The Journals of Sylvia Plath*. London: Faber & Faber, 2001.

Procter, M., Yeo, P., Lack, A. *The Natural History of Pollination*. London: HarperCollins, 1996.

Ramírez, J. *The Beehive Metaphor*. London: Reaktion Books, 2000.

Ransome, H. *The Sacred Bee in Ancient Times and Folklore*. London: BBNO, 1937, reprinted 1986.

Root, A.I. *The ABC and XYZ of Bee Culture*. Medina, OH: A.I. Root, 1877.

Roud, S. *The Penguin Book of Superstitions of Britain and Ireland*. London: Penguin, 2003.

Schaller, G.B. *The Year of the Gorilla*. London: Collins, 1965.

Schierbeek, A. *Jan Swammerdam*. Amsterdam: Swets & Zeitlinger, 1967.

Slater, L.G. *Hunting the Wild Honey Bee*. Lilliwaup, WA: Terry Publishing Co., 1969.

Steiner, R. *Bees (with an afterword on the art of Joseph Beuys by Adams, D.)*. Great Barrington, MA: Anthroposophic Press, 1998.

Stevenson, A. *Bitter Fame, A Life of Sylvia Path*. London: Viking, 1989.

Stow, G. *The Native Races of South Africa*. London: Swan Sonnenschein & Co., 1905.

Swammerdam, J. *The letter of Jan Swammerdam to Melchisedec Thévenot, with a biographical sketch by G.A. Lindeboom*. Amsterdam: Swets & Zeitlinger, 1975.

Style, S. *Honey, from Hive to Honeypot*. London: Pavilion, 1992.

Teale, E. *The Golden Throng*. Sherborne, Dorset: Alphabooks, 1981.

Toussant-Samat, M. *A History of Food*. Oxford: Blackwell, 1992.

Turnbull, C.M. *The Forest People*. London: Jonathan Cape, 1966.

Valli, E., Summers, D. *Honey Hunters of Nepal*. London: Thames & Hudson, 1988.

Van der Post, L. *The Lost World of the Kalahari*. London: Hogarth Press, 1958.

Van Lawick-Goodall, J. *In the Shadow of Man*. London: Collins, 1971.

Von Frisch, K. *Bees: Their Vision, Chemical Senses, and Language.* Ithaca, NY: Cornell University Press, 1950.

Wagner, E. *Ariel's Gift.* London: Faber & Faber, 2000.

Weiss, K. *The Little Book of Bees.* New York: Springer-Verlag, 2002.

Whynott, D. *Following the Bloom.* Harrisburg, PA: Stackpole Books, 1991.

Wildman, D. *A Compleat Guide for the Management of Bees,* 1773.

Wildman, T. *A Treatise on the Management of Bees.* Bath, Somerset: Kingsmead, 1768, reprinted 1970.

Wilkins, J., Hill, S. *The Life of Luxury.* Totnes, Devon: Prospect Books, 1994.

Wilson, E.O. *The Diversity of Life.* Cambridge, MA: Harvard University Press, 1992.

Winston, M. *Killer Bees.* Cambridge, MA: Harvard University Press, 1992.

Winston, M. *The Biology of the Honey Bee.* Cambridge, MA: Harvard University Press, 1987.

ILLUSTRATION ACKNOWLEDGMENTS

PAGE 28: © Dorothy Hodges, reproduced by kind permission of the International Bee Research Association

PAGE 30: E. H. Taylor, *Bees for Beginners* (Welwyn, England: E. H. Taylor Ltd, n.d.)

PAGE 38: © E. Hernández-Pacheco

PAGE 53: Norman de Garis Davies, *The Tomb of Rekh-mi-Re at Thebes* (New York: The Metropolitan Museum of Art Egyptian Expedition, 1943)

PAGE 58: Margaret A. Murray, *Saqqara Mastabas,* Part I (London: Bernard Quaritch, 1905) and Aylward M. Blackman and Michael R. Apted, *The Rock Tombs of Meir, Vol. 1* (London: Egyptian Exploration Society, 1914)

PAGE 62: British Museum (B117)

PAGE 66: Virgil, *Georgics,* translated by John Dryden, 1697

PAGE 78: J. G. Krünitz, *Das Wesentlichste der Bienen-Geschichte und Bienen-Zucht,* 1774

PAGE 91: Olaus Magnus, *Historia de gentibus septentrionalibus,* 1555

PAGE 104: Matteo Greuher, 1625

PAGE 106: Jan Swammerdam, *Biblia naturae,* 1737

PAGE 109: René Antoine Ferchault de Réaumur, *Mémoires pour servir à l'histoire des insectes, Vol. 5,* 1740

PAGE 112: François Huber, *New Observations upon Bees* (Illinois: American Bee Journal, 1926)

PAGE 123: Paul Dudley, *An account of a method lately found out in New England,* 1721

PAGE 130: William C. Cotton, *My Bee Book,* 1842

PAGE 137: © Mary Evans Picture Library

PAGES 144 AND 148: Lorenzo L. Langstroth, *The Hive and the Honey-Bee,* 1853 (photographer unknown)

PAGE 178: © 2004 Buckfast Abbey

PAGE 188: © Hulton Archive/Getty Images

PAGES 213 AND 215: © Hattie Ellis

PAGE 219: © Marc Renaud

INDEX

Page numbers in *italics* refer to illustrations

Napoleonic Wars, 154
Narbonne, 214
National Beekeeping Service (Spain), 166
Native Americans, 12, 120, 121, 122
nectar, 6–8, 19
    and bees' communication, 187–88
    collection by bee, 15–16
    production of wax, 86
nectaries, 15
Neolithic, 49
Nepal, 26, 48
Netherlands, 84, 102–103
Neuserre, King of Egypt, 51
New England, 121
New Forest, 80
New South Wales, 131
New York City, 17, 152, 153, 154, 217–18,
    219–22, *219*
New York State, 222
New Zealand, 129, 130–33, 197–201, 223
Newbury, Massachusetts, 120
Nile, River, 50, 51, 53, 58
Norse myths, 92
North Africa, 179, 180
North America, 25, 119–29, 184, 190–92
North American Bee Association, 146
North American Beekeepers' Society, 147
Northeastern Beekeepers' Association, 146
Northumberland, 5, 8

observation hives, 12, 69, 96–99, *98*, 102, 108,
    109, 167–68, 215–17, *215*
Odin, 92
Odysseus, 74
Oertel, Dr. Everett, 122
oilseed rape, 7, 204–205
Old Testament, 65, 127
Olympus, Mount, 61, 67
Opie, Iona, *Dictionary of Superstitions*, 137
orchids, 23–24
organophosphates, 225, 226
Otis, R.C., 146
"ox-born bees," 64–65, *66*, 129
Oxford University, 98, 99

Pagliaro, Paolo, 71–73
Paine, Tom, 205
paintings, rock art, 37–42, *38*, 46–47, 165
Palacio Güell, Barcelona, 167
Palladius, 70
Panama, 152
Pantalica, 73–74
parabolic arches, 166–67
parasites, 223–24, 225–26
parc Georges Brassens, Paris, 162
Paris, 161–62, 168, 209–16

Parsons, S.B., 154
Patrick, St., 79
Patroclus, 68
Pellett, Frank, 120
    History of American Beekeeping, 147
Pennsylvania, 121, 151–52
Perdita minima, 25
performance art, 164–66
Petrified Forest National Park, 20
Pettigrew, *The Handy Book of Bees*,
    156–57
Philadelphia Inquirer, 191
Philippines, 63–64
Pilgrim Fathers, 120–21
place names, Wisconsin, 122
Plath, Otto, 170–71
Plath, Sylvia, 170–73
    Ariel, 173
Pliny the Elder, 69, 87, 92
poetry, 63–64, 67, 113–15, 171, 172–73, 207,
    228
Poland, 78–79
Pollard, Charlie, 172
pollen: colors, 28
    medicinal uses, 194, 195, 201
    migratory beekeeping, 53
    pollination, 18–19, 24, 226
pollen baskets, 28, *28*
Potter, Beatrix, 81–82
Prairie Farmer, 150
prehistoric man, 35, 37–42
propolis, 69, 143, 193, 194, 228
Provence, 179
Purbeck, Isle of, 202
Pyrenees, 209, 214
Pythagoras, 72

queen cells, 30–31
queens, 29–30, *30*
    artificial breeding, 177–79, 180–81, 225
    egg-laying, 30–31
    gender, 113
    hatching, 30
    mating, 30, 31, 110–12, 160
    "queen-excluder," 149
    royal jelly, 195
    stings, 30
    swarming, 29–30
    virgin flight, 111, 160
Quinby, Moses, 149

Ra, 58
Ramses III, Pharaoh, 51
Ramírez, Juan Antonio, *The Beehive
    Metaphor*, 165–168, 169
Ramírez de la Morena, Lucio, 165